MATHEMATICAL JOURNALS
An Annotated Guide

compiled by
DIANA F. LIANG

The Scarecrow Press, Inc.
Metuchen, N.J., & London
1992

British Library Cataloguing-in-Publication data available

Library of Congress Cataloging-in-Publication Data

Liang, Diana F.
    Mathematical journals : an annotated guide / compiled by Diana
F. Liang.
        p.   cm.
    Includes index.
    ISBN 0-8108-2585-6 (alk. paper)
    1. Mathematics--Periodicals--Bibliography. I. Title.
Z6653.L5 1992
 [QA1]
016.51'05--dc20   92-18459

TO

JOSEPH, my husband

and sons
PATRICK   and   MICHAEL

TABLE OF CONTENTS

PREFACE

A mathematician's laboratory equipment is the journals used for research. These journals become an essential tool, a necessary and important factor in producing the results of their studies. In addition, these journals act as vehicles for mathematicians to share and promote the most recent developments in their fields. Since mathematics is the basic background for all disciplines of science, journals provide a medium which promotes the exchange of ideas and discussion among all scientists.

Mathematical Journals assembles a group of approximately 350 active English-language journal titles from thousands of titles currently in print. The purpose of this publication is three-fold: (1) to provide an overview of living English-language publications in the field of mathematics; (2) to assist mathematicians and other scientists by providing them with a choice of journals to use for reading or for the submission of manuscripts, and (3) to raise librarians' awareness and assist them in selecting and evaluating their own collections.

Coverage of titles in the *Guide* is current; all the titles were still in publication during 1991. Mathematics, statistics and mathematics-related computer science titles comprise most of the entries. The compiler selected journals published primarily in English. This includes those printed in English but from non-English-speaking nations. The titles were selected through database searches, reviews of current publications, and titles covered in the *Mathematical Review*. The compiler began the search by personally examining each title from the journal collection at the University of South Florida Library. The research was continued at the University of California, Berkeley Mathematics Library, the Library of Congress, and the University of Florida Science Library. Other bibliographic tools, such as *Ulrich's Plus, International Periodical Directory, Faxon Librarians' Guide to Serials 1991, UMI Catalog,* and the databases of NOTIS and OCLC were also consulted. When all the information online was coupled with the compiler's cataloging experience, the search for the history of each periodical turned out to be more intelligible and the compilation of the work progressed smoothly.

Many thanks go to Dr. Greg McColm for his generous assistance and advice and to Ms. Nancy Jacobs for inputting the material into the computer and coming back with the final formatted manuscript. Finally, I would like to thank my family members for their support and encouragement. Without their understanding, this project could never have been accomplished.

Despite my best intentions and efforts, errors still exist and titles to be covered are far from complete. I would appreciate any kind of suggestions and corrections. I'd like to have this guide become a valuable tool and dedicate it to all mathematicians and everyone else who is interested in mathematics.

February 14, 1992

Diana F. Liang
Tampa, Florida

INTRODUCTION

Mathematical Journals is an annotated bibliography intending to cover all the living English-language mathematical journals. Titles in this book are arranged alphabetically with each entry having two parts. The first part provides bibliographic information for the title and the second part is a descriptive annotation. The information contained in the first part includes:

TITLE: Any title with an asterisk (*) refers to a journal not physically examined by the compiler. Information was gathered from a publisher's catalog or reference tool.

ISSN (International Standard Serial Number): This international numerical code identifies concisely, uniquely, and unambiguously a serial publication. The ISSN usually appears on the cover of a serial. The ISSNs of previous titles are also included in the entry.

HISTORY: A detailed history of the journal publication. Covers not only the record of the variant title changes but the period covered for that particular title.

EDITOR: Person in charge of the publication and its contents.

PUBLISHER: Agency responsible for the publication of the journal. A directory of all publishers mentioned in this book is provided in the back. For some foreign publishers, the distribution agency is included.

DATE FOUNDED: Publication date of the current title. Information was supplied either by the publisher, cataloging source or other reference tool. The most recent publication date is used based on the history of a title which has undergone a merger or name change(s).

FREQUENCY: Indicates how often the journal is published. Often this information is contained in the issue.

PRICE: Amount listed reflects the 1991 annual subscription price in US dollars. Usually the information is also listed in the publication or the publisher's catalog. Due to the dollar rate change, the price listed in this bibliography should only be used as a guide.

MICROFORM: Availability of the microform format and where it can be purchased. The address of the vendors appears in the Directory of Publishers and Distributors.

LANGUAGE: Primarily English. Other languages may be listed as an aid for submission.

REPRINTS: If reprint services are available, the information is included in the Bibliography and the address of the source would be listed in the Directory of Publishers and Distributors.

CIRCULATION: This information can be used as a guide.

INDEXED/ABSTRACTED: Contained in this area are the indexes and abstracts in which the journal is indexed or abstracted. Not all indexes and abstracts are listed. Only the most relevant ones are included. A Table of Abstracts and Indexes is in this book.

TARGET READER: Based on the nature of the contents of the journal, the interested readers are mostly suggested by the compiler, some are suggested by the publisher's catalog.

ABSTRACTS OF PAPERS PRESENTED TO THE AMERICAN MATHEMATICAL SOCIETY
ISSN: 0192-5857

HISTORY: Between 1894 and 1963, abstracts of contributed and invited talks presented at AMS meetings appeared in the *Bulletin of the AMS* (Vol 1-57). Between 1964 and 1979, they were printed in the *Notices of the Arts.*

PUBLISHER: American Mathematical Society
DATE FOUNDED: 1979   FREQUENCY: Bi-Monthly
PRICE: $54   LANGUAGE: English
          (Inst Memb) $43   MICROFORM: UMI
          (Indv Memb) $32   CIRC: 4400
INDEXED/ABSTRACTED: Math R
TARGET READER: Mathematicians in all fields

Contains abstracts of invited hour addresses, of papers presented in special sessions or in sessions for contributed papers, and of papers presented to the Society.

ACTA APPLICANDAE MATHEMATICAE
ISSN: 0167-8019

EDITOR: Michiel Hazewinkle
                              Center for Math
                              and Computer Science
                              P. O. Box 4079
                              1009 AB Amsterdam
                              THE NETHERLANDS
MS REQUIREMENT: Triplicate
PUBLISHER: Kluwer Academic Publishers
DATE FOUNDED: 1983   FREQUENCY: Monthly
PRICE: $518   LANGUAGE: English
MICROFORM: UMI   CIRC: 500
INDEXED/ABSTRACTED: Appl Mech Rev, Curr Cont, ISI/Compumath, Math R, Ref Zh, Zent Math
TARGET READER: Mathematicians in all fields

Devoted to the art and the techniques of applying mathematics and the development of new applicable mathematical theories. It contains papers on the different aspects of the relation between theory and application. Special forum introduces mathematicians and other scientists to new techniques of applying mathematics to new fields.

ACTA ARITHMETICA
ISSN: 0065-1036

PUBLISHER: Polish Academy of Sciences
DATE FOUNDED: 1936 FREQUENCY: Irr
PRICE: Request from LANGUAGE: English, French,
                    the Publisher   German
INDEXED/ABSTRACTED: Math R
TARGET READER: Number Theorists

Publishes research papers in the field of number theory.

ACTA INFORMATICA
ISSN: 0001-5903

EDITOR: F. L. Bauer
PUBLISHER: Springer-Verlag
DATE FOUNDED: 1971   FREQUENCY: 8/Yr.
PRICE: $574.33 MICROFORM: UMI
REPRINTS: ISI   LANGUAGE: Multilingual
                           (Mainly in English)
INDEXED/ABSTRACTED: Compumath, Curr Cont, Comput Rev,
Eng Ind, Ind Sci Rev, Math R, Zent Math
TARGET READER: Theoretical Computer Scientists, Software
Engineers, Programmers

Provides international dissemination of contributions on the art, discipline and science of informatics. Its scope covers design, description, presentation, and analysis of programs, information structures, computing systems and the interaction between components thereof.

ACTA MATHEMATICA
ISSN: 0001-5962

EDITOR:	Lennart Carleson
MS REQUIREMENT:	Duplicate, dbl spaced
SEND MS TO:	Acta Mathematica
	Institut Mittag-Leffler, Auravägen 17,
	S-18262 Djursholm, SWEDEN
PUBLISHER:	Institut Mittag-Leffler
DATE FOUNDED:	1882 FREQUENCY: Quarterly
PRICE:	$200 LANGUAGE: English, French,
	German

INDEXED/ABSTRACTED: Compumath, Ind Sci Rev, Math R, Sci Cit Ind
TARGET READER: Mathematicians in all fields

Publishes original mathematical research papers.

ACTA MATHEMATICA HUNGARICA
ISSN: 0236-5294

HISTORY: *Hungarica Acta Mathematica*, V1#1-V1#4
(1946-1949)
Acta Mathematica Academiae Scientiarum Hungaricae, [ISSN: 0001-5954] T1-T40
(1950-1982)

EDITOR:	K.Tandori
	Acta Mathematica
	P. O. Box 127
	H-1364 Budapest
MS REQUIREMENT:	One copy, no abstract
PUBLISHER:	Akademiai Kiado
	(Publishing House of the

Hungarian Academy of Scienes)

SUBSCRIPTION:	Kultura Foreign Trading Co.
	P. O. Box 149
	H-1389 Budapest, HUNGARY
DATE FOUNDED:	1983 FREQUENCY: Quarterly
PRICE:	$60 CIRC: 1000
LANGUAGE:	English, German, French, Russian

INDEXED/ABSTRACTED: Compumath, Curr Cont, Ind Sci Rev, Math R, Sci Cit Ind

TARGET READER: Pure and Applied Mathematicians

Covers a wide scope in the field of mathematics. It comprises theory of sets, mathematical logic, classical and modern analysis, algebra, number theory, geometry, topology, combinatorics, mathematical statistics, probability theory as well as information theory.

ACTA MATHEMATICA SINICA
ISSN: 1000-9574

EDITOR:	Wang Juan
	Institute of Math
	Academia Sinica
PUBLISHER:	Science Press, Beijing
	(outside China)
	VSP/Int'l Science Publishers
DATE FOUNDED:	1985 FREQUENCY: Quarterly
PRICE:	$258.74 LANGUAGE English
CIRC:	6000

INDEXED/ABSTRACTED: Math R

TARGET READER: Pure and Applied Mathematicians

Publishes English papers from all branches of both pure and applied mathematics. It covers fields such as algebra, analysis, geometry, topology, number theory, numerical analysis, probability and statistics, and applied mathematics.

ACTA SCIENTIARUM MATHEMATICARUM
ISSN: 0001-6969

HISTORY: *Acta Litterarum ac Scientiarum Regiae Universitatis Hungaricae Francisco-Josephinae: Sectio Scientiarum Mathematicarum* [ISSN: 0324-5462] T1-T9 (1922-1940)

MANAGING EDITOR: Josef Merza
H-1053 Budapest
Realtanoda, U 13-15
HUNGARY
MS REQUIREMENT: Duplicate
PUBLISHER: Kultura Foreign Trading Co.

		FREQUENCY:	
DATE FOUNDED:	1941	FREQUENCY:	Quarterly
PRICE:	$52	MICROFORM:	UMI
CIRC:	1200	LANGUAGE:	English, French, German

INDEXED/ABSTRACTED: Math R, Sci Cit Ind
TARGET READER: Mathematicians in all fields

Publishes original research papers on mathematics and its diverse fields of application in science, technology, economics, etc.

ADVANCES IN APPLIED MATHEMATICS
ISSN: 0196-8858

EDITOR: Gian-Carlo Rota
MIT
Cambridge, MA 02139
MS REQIREMENTS: Original and one copy, prefer triple spaced, 50-150 word abstract (optional)
PUBLISHER: Academic Press

DATE FOUNDED:	1980	FREQUENCY:	Quarterly
PRICE:	$126	LANGUAGE:	English

INDEXED/ABSTRACTED: Compumath, Ind Sci Rev, Math R
TARGET READER: Applied Mathematicians, Computer Scientists, Statisticians

Publishes papers in all areas of applied mathematics. Particular regard will be given to papers that represent a substantial advance in their field. Well written expository surveys are also published. Features articles on discrete applied mathematics, statistics, mathematical biology, dynamical systems, algorithms, experimental mathematics, and theoretical computer science.

ADVANCES IN APPLIED PROBABILITY
ISSN: 0001-8678

EDITOR:	C. C. Heyde
	Australian National University
SEND MS TO:	The Executive Editor
	Applied Probability
	Department of Probability/Statistics
	The University
	Sheffield S3 7RH, ENGLAND

MS REQUIREMENT: Triplicate, dbl spaced, 4-10 line abstract, 1991 Mathematics Subject Classification numbers, list of key words
PUBLISHER: Applied Probability
DATE FOUNDED: 1969   FREQUENCY: Quarterly
PRICE: (Inst) $150   CIRC: 1100
                    (Indv) $50   LANGUAGE: English, French
INDEXED/ABSTRACTED: Compumath, Curr Cont, Ind Sci Rev, Math R, Sci Abstr, Sci Cit Ind
TARGET READER: Applied Mathematicians, Probablists, Statisticians

Publishes review papers summarizing and coordinating recent results in any of the fields of applied probability. It also publishes longer research papers in applied probability, expository papers on branches of mathematics of interest to probabilists, and papers outlining areas in the biological, physical, social, and technological sciences in which probability models can be usefully developed.

ADVANCES IN MATHEMATICS
ISSN: 0001-8708

EDITOR: Gian-Carlo Rota
                            MIT
                            Cambridge, MA   02139
MS REQUIREMENT: Duplicate, dbl spaced, 50-150 word
abstract (optional)
PUBLISHER: Academic Press
DATE FOUNDED: 1965   FREQUENCY: Monthly
PRICE: $636   LANGUAGE: English
INDEXED/ABSTRACTED: Compumath, Curr Cont, Ind Sci Rev,
Math R, Sci Cit Ind
TARGET READER: Researchers interested in all areas of pure
mathematics

Publishes papers in pure mathematics, and provides research
mathematicians with an effective medium for communicating important
recent developments in their areas of specialization to colleagues and to
scientists in related disciplines.   Considered to be one of the top
journals in the field.

AEQUATIONES MATHEMATICAE
ISSN: 0001-9054

EDITOR: Prof. J. Aczél
                            University of Waterloo
                            Waterloo, Ontario
                            N2L 3G1   CANADA
MS REQUIREMENT: Original and two copies, AMS
Classification Numbers,maximum 25 lines of summary.
PUBLISHER: Birkhaüser Verlag
DATE FOUNDED: 1968   FREQUENCY: 6/Yr.
PRICE: $300   MICROFORM: UMI
LANGUGE: English, French, German
INDEXED/ABSTRACTED: Math R, Ref Zh
TARGET READER: Mathematicians in all fields

Publishes papers in pure and applied mathematics in general, but, in particular, papers on functional equations, combinatorics, dynamical systems, and numerical analysis.

ALGEBRA AND LOGIC
ISSN: 0002-5232

English Translation of: *Algebra i Logika* [ISSN: 0373-9252]

EDITOR:	Yu. L. Ershov
PUBLISHER:	Consultants Bureau
	Plenum Press

DATE FOUNDED: 1968   FREQUENCY: Bi-Monthly
PRICE: $635 LANGUAGE: English
INDEXED/ABSTRACTED: Comput Inf Syst Abstr J, Math R, Zent
Math TARGET READER: Algebraists, Logicians

Reports results of the latest research in the areas of modern general algebra and logic considered primarily from an algebraic viewpoint.

ALGEBRA UNIVERSALIS
ISSN: 0002-5240

EDITOR:	G. Grätzer
	Department of Math
	University of Manitoba
	Winnipeg R3T 2N2, CANADA
PUBLISHER:	Birkhaüser Verlag

DATE FOUNDED: 1968   FREQUENCY: Quarterly
PRICE: $236 LANGUAGE: English
INDEXED/ABSTRACTED: Math R, Ref Zh
TARGET READER: Algebraists

Publishes papers on the applications of algebra to other fields.

ALGEBRAS, GROUPS, AND GEOMETRIES
ISSN: 0774-9937

EDITORIAL   495 A-19
OFFICES: P. O. Box 1577
                     Palm Harbor, FL 34682-1577
MS REQUIREMENT: Duplicate, AMS Subject Classification
Numbers, accept both typed or typeset form
PUBLISHER: Hardronic Press
DATE FOUNDED: 1985 FREQUENCY: Quarterly
PRICE: $150 CIRC: 1000
INDEXED/ABSTRACTED: Chem Abstr, Curr Cont, Math R, Sci
Abstr, Sci Cit Ind
TARGET READER: Algebraists, Geometricians

Publishes original research papers, research-expository and survey articles in linear algebra, Lie groups, and their isotopics, non-associative rings and algebras, linear and multilinear algebras, differential geometries.

AMERICAN JOURNAL OF MATHEMATICAL AND MANAGEMENT SCIENCES
ISSN: 0196-6324

EDITOR: Edward J. Dudewicz
                     Department of Math
                     Syracuse University
                     Syracuse, NY 13244
MS REQUIREMENT: Five copies
PUBLISHER: American Science Press
DATE FOUNDED: 1981   FREQUENCY: Quarterly
PRICE: $174.50 LANGUAGE: English
Back Issues Available   CIRC: 1000
INDEXED/ABSTRACTED: Curr Cont, Curr Ind Stat, Math R, Oper
Res Manag Sci, Phys Abstr, SSCI
TARGET READER: Mathematicians in various fields, Management
Scientists, Operation Researchers, Computer Scientists

Focuses on new work in the various areas of the mathematical and management sciences including statistics, probability, decision theory, information theory, engineering systems, expert systems, and artificial intelligence.

AMERICAN JOURNAL OF MATHEMATICS
ISSN: 0002-9327

EDITOR:	Jun-Ichi Igusa
	Department of Math
	Johns Hopkins University
	Baltimore, MD 21218
MS REQUIREMENT:	Duplicate, dbl spaced, introduction
PUBLISHER:	Johns Hopkins University Press
DATE FOUNDED:	1878 FREQUENCY: Bi-Monthly
PRICE: (Inst)	$146 MICROFORM: UMI
(Indv)	$49 CIRC: 1900
LANGUAGE:	English, French, German, Italian, Spanish

INDEXED/ABSTRACTED: Compumath, Curr Cont, Ind Sci Rev, Math R, Sci Cit Ind
TARGET READER: Mathematicians in all fields, Librarians

Presents pioneering work in applied and pure mathematics. An essential resource for students and scholars. Used as a basic reference work in academic libraries.

AMERICAN MATHEMATICAL MONTHLY
ISSN: 0002-9890

EDITOR:	John Ewing
	Department of Math
	Indiana University
	Bloomington, IN 47405
MS REQUIREMENT:	Three copies, dbl spaced, Mathematics

Subject Classification Numbers
PUBLISHER: Mathematical Association of America
DATE FOUNDED: 1894 FREQUENCY: 10/yr

PRICE: $128   LANGUAGE: English
                    (Memb) $32   MICROFORM: UMI
CIRC: 20,000
INDEXED/ABSTRACTED: Compumath, Curr Cont, Gen Sci Ind,
Ind Sci Rev, Math R, Sci Cit Ind
TARGET READER: Mathematicians in all fields, Mathematics
Educators

Includes expository articles on all components of mathematics, pure and
applied, old and new, with regular columns devoted to basic and
complex problems and reviews. It also includes mathematical and
classroom notes, a section on the teaching of mathematics, elementary
and advanced problems and telegraphic and extended book and film
reviews.

AMERICAN STATISTICIAN
ISSN: 0003-1305

HISTORY: *American Statistical Association Bulletin*, V1-V7
                (1938-1947)

EDITOR: William R. Schucany
                Department of Statistical Science
                Southern Methodist University
                Dallas, TX 75275-0332
MS REQUIREMENT: Four copies, dbl spaced
PUBLISHER: American Statistical Association
DATE FOUNDED: 1947 FREQUENCY: Quarterly
PRICE: (Inst) $225 MICROFORM: UMI, MIM
                (Memb) $ 85 LANGUAGE: English
                (Stud) $ 32 PUBLICATION CHARGES:
                (Corp) $480 to the institution or granting
                             agency
INDEXED/ABSTRACTED: Chem Abstr, Compumath, Math R,
SSCI, Sci Abstr, Sci Cit Ind, Soc Sci Ind
TARGET READER: Statisticians, Computer Scientists

Publishes papers for the accent on teaching materials, statistical computing software reviews, and new developments in statistical computing.

ANALYSIS (Munich, Germany)
ISSN: 0174-4747

HISTORY: *Analysis* (Wiesbaden) [ISSN: 0174-4747] V1
                (1981)

EDITOR: A. M. Garsia
                               Department of Math
                               University of California/San Diego
                               La Jolla, CA 92093
MS REQUIREMENT: Maximum 10 line abstract in English,
AMS 1980 Classification Numbers
PUBLISHER: R. Oldenbourg Verlag GmbH
DATE FOUNDED: 1982   FREQUENCY: Quarterly
PRICE: $187.41 CIRC: 250
LANGUAGE: English
INDEXED/ABSTRACTED: Math R
TARGET READER: Mathematicians interested in the field of
analysis

Publishes original research papers from the field of analysis, in particular, classical analysis and its applications. Papers on analytic number theory and survey articles are included.

ANALYSIS MATHEMATICA
ISSN: 0133-3852

EDITORIAL   Bolyai Institute
BOARD: Aradi Vertanuk tere 1
                        6720 Szeged, HUNGARY
MS REQUIREMENT: Duplicate, English abstract for Russian
papers, Russian Abstract for papers of other languages, abstract
limited to 20 lines.
PUBLISHERS: Academy of Sciences of the USSR

Hungarian Academy of Sciences
DISTRIBUTED BY: Pergamon Press
DATE FOUNDED: 1975 FREQUENCY: Quarterly
PRICE: $230 MICROFORM: UMI
LANGUAGE: English, Russian, French, German
INDEXED/ABSTRACTED: Math R, Sci Abstr
TARGET READER: Analysts

Dedicated primarily to problems of classical mathematical analysis such as differentiation and integration of functions, measure theory, analytic and harmonic functions, fourier analysis, etc. Publishes research papers containing essential new results with complete proofs and occasionly survey papers prepared at the initiative of the Editorial Board.

*ANNALES MATHEMATICAE SILESIANAE**
ISSN: 0860-2107

HISTORY: *Prace Matematyczne* [ISSN: 0208-5410] V1-V12
                (1969-1982)

PUBLISHER: Uniwersytet Slaski w Katowicach
DISTRIBUTED BY: CHZ Ars Polona
DATE FOUNDED: 1985 FREQUENCY: Irr
PRICE: Varies LANGUAGE: English
INDEXED/ABSTRACTED: Math R
TARGET READER: Mathematicians in all fields

Covers all aspects of pure and applied mathematics in general, in particular: algebra and theory of numbers, differential equations and dynamical systems, functional and real analysis, functional equations, geometry and topology.

ANNALI DELLA SCUOLA NORMALE SUPERIORE DI PISA CLASSE DI SCIENZE
ISSN: 0391-173X

HISTORY: *Annali della R. Scuola Normale Superiore di Pisa:*
                Scienze fisiche e matematiche/V1-V10
                (1871-1908)
                Annali della R. Scuola Normale Superiore di Pisa:
                Scienze fisiche, matematiche e naturali/V11-V12
                (1910-1912)
                Annali della R. Scuola Normale Superiore di Pisa:
                Scienze fisiche e matematiche/V13 (1919)
                Annali della R. Scuola Normale Universitaria di
                Pisa: Scienze fisiche e matematiche/V14-V15
                (1922-1927)
                Annali della R. Scuola Normale Superiore di Pisa:
                Classe di Scienze fisiche, matematiche, e Naturali
                V16 (1929-1930)
                Annali della R. Scuola Normale Superiore di Pisa:
                Scienze fisiche e matematiche/Series 2, V1-V12
                (1932-1943)
                Annali della Scuola Normale Superiore di Pisa:
                Scienze fisiche e matematiche/V12 #3&4, Series 3,
                V27 (1943-1973)

EDITOR: Edoardo Vesentini
SEND MS TO: Amministrazione Degli Annali
                        Classe di Scienze
                        Scuola Normale Superiore
                        56100 Pisa, ITALY
MS REQUIREMENT: Duplicate, no abstract
PUBLISHER: Amministrazione Degli Annali
DATE FOUNDED: 1974   FREQUENCY: Quarterly
PRICE: $130   LANGUAGE: English, French,
                                             German, Italian
INDEXED/ABSTRACTED: Math R
TARGET READER: Mathematicians in all fields.

Publishes longer and expository mathematical papers in all areas.

ANNALS OF GLOBAL ANALYSIS AND GEOMETRY
ISSN: 0232-704X

EDITORIAL   Fachbereich Mathematik
OFFICE: der Humboldt Universitat
                       Postfach 1297, O-1086
                       Berlin, GERMANY
MS REQUIREMENT: Original and one copy, dbl spaced, abstract
and bibliography
PUBLISHER: Kluwer Academic Publishers
DATE FOUNDED: 1983   FREQUENCY: 3/yr.
PRICE: $114   LANGUAGE: English
INDEXED/ABSTRACTED: Math R
TARGET READER: Analysts, Theoretical Physicists,
Mathematicians interested in global problems of geometry and
analysis

Publishes papers to contribute to an enlargement of the international
exchange of research results in the fields of Global Analysis and
Geometry. It treats in particular global problems of geometry and
analysis as well as the interactions between these fields and their
application to problems of Theoretical Physics.

ANNALS OF MATHEMATICS
ISSN: 0003-486X

HISTORY: *Analyst* [ISSN: 0741-7918] V1-V10 (1874-1883)

SEND MS TO: Katharine Carter
                       Annals of Mathematics
                       Fine Hall/Washington Road
                       Princeton, NJ 08544
MS REQUIREMENT: Duplicate, dbl spaced
PUBLISHER: Johns Hopkins University Press
DATE FOUNDED: 1884   FREQUENCY: Bi-Monthly
PRICE: (Inst) $180   MICROFORM: UMI
              (Indv) $ 60   CIRC: 2500
LANGUAGE: English, French, German
INDEXED/ABSTRACTED: Compumath, Curr Cont, Ind Sci Rev,
Math R
TARGET READER: Mathematicians interested in pure mathematics.

Publishes research papers in pure mathematics.

ANNALS OF MATHEMATICS AND ARTIFICIAL INTELLIGENCE
ISSN: 1012-2443

EDITOR: Martin Charles Golumbic
                                 IBM Israel Scientific Center
                                 Technion City
                                 Haifa, ISRAEL
PUBLISHER: J. C. Baltzer, AG
DATE FOUNDED: 1990   FREQUENCY: Irr
PRICE: (Inst) $221   LANGUAGE: English
                    (Indv) $92
INDEXED/ABSTRACTED: Comput Control Abstr, Elect Electron
Abstr, Phys Abstr
TARGET READER: Theoretical Computer Scientists

Reports significant contributions on the interaction of mathematical and computational techniques reflecting the evolving disciplines of artificial intelligence.

ANNALS OF PROBABILITY
ISSN: 0091-1798

HISTORY: *Annals of Mathematical Statistics*
                      [ISSN: 0003-4851] V1-V43 (1930-1972)

EDITOR: Burgess Davis
                      Department of Math/Statistics
                      Purdue University
                      West Lafayette, IN 47907
MS REQUIREMENT: Triplicate, dbl spaced, AMS 1980 Subject
Classification Numbers, key words/phrases, summary not to exceed
150 words.

PUBLISHER: Institute of Mathematical Statistics
DATE FOUNDED: 1973   FREQUENCY: Quarterly
PRICE: (Inst) $150   MICROFILM: UMI
                    (Memb) $ 60   CIRC: 3100
LANGUAGE: English
INDEXED/ABSTRACTED: Compumath, Curr Cont, Ind Sci Rev,
Math R, Sci Cit Ind
TARGET READER: Probabilists, Statisticians

Publishes contributions to the theory of probability and statistics and
their application.

ANNALS OF PURE AND APPLIED LOGIC
ISSN: 0168-0072

HISTORY: *Annals of Mathematical Logic* [ISSN: 0003-4843]
                V1-V23 (1970-1982)

EDITOR: Y. Gurevich
                        Department of Computer Science
                        and Engineering
                        University of Michigan
                        Ann Arbor, MI 48109-1003
PUBLISHER: Elsevier
DATE FOUNDED: 1983   FREQUENCY: Semi-Monthly
PRICE: $898.87 LANGUAGE: English
CIRC: 700   MICROFORM: RPI, UMI
INDEXED/ABSTRACTED: Compumath, Curr Cont, Math R, Phil
Ind, Sci Abstr
TARGET READER: Logicians, Set Theorists, Set Theoretic
Topologists, Researchers in Automata Theory, Computer Scientists

Publishes longer papers (25 pages and more) on topics in pure and
applied logic.   The foundations of mathematics and the area of
theoretical computer sciences and other disciplines which are of direct
interest to mathematical logic.

ANNALS OF STATISTICS
ISSN: 0090-5364

HISTORY: *Annals of Mathematical Statistics*
                [ISSN: 0003-4851] V1-V43 (1930-1972)

EDITOR: Arthur Cohen
                       Department of Statistics
                       Rutgers University/Busch Campus
                       New Brunswick, NJ 08903
MS REQUIREMENT: Four copies, dbl spaced, AMS 1980
Subject Classification Numbers, summary under 150 words.
PUBLISHER: Institute of Mathematical Statistics
DATE FOUNDED: 1973 FREQUENCY: Quarterly
PRICE: (Inst) $110 MICROFORM: UMI
              (Indv) $ 60 CIRC: 4600
LANGUAGE: English
INDEXED/ABSTRACTED: Compumath, Curr Cont, Ind Sci Rev,
Math R, SSCI, Sci Cit Ind
TARGET READER: Statisticians

Publishes significant contributions to the theory of statistics and its
applications. The emphasis of the publication is on importance and
interest. Especially appropriate are authoritative expository papers and
surveys or areas in vigorous development.

ANNALS OF THE INSTITUTE OF STATISTICAL MATHEMATICS
ISSN: 0020-3157

EDITOR: H. Akaike
                       AISM Editorial Office
                       Institute of Statistical Mathematics
                       4-6-7 Minami-Azabu
                       Minato-ku, Tokyo 106, JAPAN
MS REQUIREMENT: Four copies, dbl or triple spaced, maximum
150 word abstract, ten key word phrases
PUBLISHER: Kluwer Academic Publishers

DATE FOUNDED: 1949 FREQUENCY: Quarterly
PRICE: $213 CIRC: 1500
LANGUAGE: English
INDEXED/ABSTRACTED: Compumath, Jap Per Ind, Math R, Sci Cit Ind
TARGET READER: Statisticians and research workers working with the common purpose of advancing human knowledge through the development of the science and technology of statistics.

Publishes broad coverage of statistical papers. The emphasis of papers is: a) establishment of new areas of application, b) development of new procedures and algorithms, c) development of unifying theories, d) analysis and improvement of existing procedures and theories, e) communication of empirical findings with supporting real data.

APPLICABLE ALGEBRA IN ENGINEERING, COMMUNICATIONS, AND COMPUTING*

ISSN: 0938-1279

EDITOR: Jacques Calmet
PUBLISHER: Springer-Verlag
DATE FOUNDED: 1990 FREQUENCY: 4/Yr.
PRICE: $137 LANGUAGE: English
INDEXED/ABSTRACTED: Comput Control Abstr, Elect Electron Abstr, Phys Abstr
TARGET READER: Algebraists, Computer Scientists, Applied Mathematicians

Publishes mathematical research papers on algebraic methods and techniques relevant to all domains of computers, intelligent systems and communications.

APPLICABLE ANALYSIS
ISSN: 0003-6811

EDITOR: Robert P. Gilbert

Department of Math and
Center for Computational Math
University of Delaware
Wilmington, DE 19711

MS REQUIRMENT: Duplicate, dbl spaced, 300 word summary,
100-150 word abstract, six key words
PUBLISHER: Gordon & Breach Science Publishers
DATE FOUNDED: 1970 FREQUENCY: Quarterly
PRICE: $1566 MICROFORM: MIM
LANGUAGE: English (preferred), French, German
INDEXED/ABSTRACTED: Appl Mech Rev, Compumath, Math R,
Sci Abstr
TARGET READER: Analysts, Engineers

Concerned with analysis that has been applied, or potentially applicable,
to the solution of scientific, technical engineering and social problems.
Aims to encourage the development of applicable analysis rather than
of generalizations merely for the purposes of abstraction.

APPLIED MATHEMATICAL MODELLING
ISSN: 0307-904X

EDITOR: Mark Cross
                     Centre for Numerical
                     Modelling and Process Analysis,
                     Thames Polytechnic
                     Wellington Street
                     London SE18 6PF, ENGLAND
MS REQUIRMENT: Triplicate, dbl spaced, 50-200 word
abstract, 4-6 key words
PUBLISHER: Butterworth Scientific Ltd.
DATE FOUNDED: 1976 FREQUENCY: Monthly
PRICE: $440 MICROFORM: UMI
LANGUAGE: English
INDEXED/ABSTRACTED: Appl Mech Rev, Compumath, Curr
Tech Ind, Eng Abstr, Math R, Sci Abstr
TARGET READER: Applied mathematicians, Computer

Scientists

Publishes original research papers approximately 6000-8000 words and invited only review articles. Welcomes contributions on all aspects of mathematical modelling pertinent to practical systems analysis.

APPLIED MATHEMATICS AND COMPUTATION
ISSN: 0096-3003

EDITOR: Melvin Scott
                             Boeing Computer Service
                             Alabama Supercomputer Network
                             Huntsville, AL 35806
MS REQUIREMENT: Original plus two copies, dbl spaced, up to
200 word abstract.
PUBLISHER: Elsevier
DATE FOUNDED: 1975   FREQUENCY: 18/yr.
PRICE: $810   MICROFORM: RPI
LANGUAGE: English
INDEXED/ABSTRACTED: Appl Mech Rev, Biol Abstr,
Compumath, Curr Cont, Eng Ind, Math R, Sci Cit Ind
TARGET READER: Applied Mathematicians, System Analysts,
Operations Researchers and Numerical Analysts

Addresses work at the interface between applied mathematics, numerical computation and applications of systems-oriented ideas to the physical, biological, social and behavioral sciences. Emphasizes work in system science, both theoretical and applied.

APPLIED MATHEMATICS AND OPTIMIZATION
ISSN: 0095-4616

EDITOR: G. Kallianpur
                             Department of Statistics
                             University of No. Carolina
                             Chapel Hill, NC 27599-3260

MS REQUIREMENT: Duplicate, abstract must be in English
PUBLISHER: Springer-Verlag
DATE FOUNDED: 1974   FREQUENCY: Bi-Monthly
PRICE: $275   MICROFORM: UMI
REPRINT: ISI   LANGUAGE: English, French
INDEXED/ABSTRACTED: Compumath, Curr Cont, Eng Ind, Ind
Sci Rev, Math R, Sci Abstr, Sci Cit Ind, Zent Math
TARGET READER: Applied Mathematicians, Researchers in
Optimization Studies

Publishes original work on mathematical problems of optimization.
Includes contributions that are complete in both theoretical and applied
aspects, papers dealing with applied mathematical topics with a practical
implication, reports concerning modeling and identification of systems,
and critical surveys of new advances in theory and application.

APPLIED MATHEMATICS LETTERS
ISSN: 0893-9659

EDITOR: Ervin Y. Rodin
                       Department of Systems Science and Math
                       Washington University
                       P. O. Box 1040
                       St. Louis, MO 63130
MS REQUIREMENT: Send to appropriate editor, paper published
within three months of acceptance, brief abstract, Triplicate (not to
exceed 4 pages) or camera ready form
PUBLISHER: Pergamon Press
DATE FOUNDED: 1988   FREQUENCY: 6/yr.
PRICE: $245   MICROFORM: UMI
LANGUAGE: English
INDEXED/ABSTRACTED: Comput Inf Syst Abstr J, Math R, Oper
Res Manag Sci
TARGET READER: Mathematicians in all fields, Computer
Scientists, Systems Scientists

Provides a means of very rapid publication for important but brief applied mathematical papers. All areas of mathematics are appropriate from number theory through Lie algebras to differential games. Accepts papers involving a novel application or utilization of mathematics, or a development in the methodology of applied mathematics.

*APPLIED MATHEMATICS NOTES**
ISSN: 0700-9224

PUBLISHER: Canadian Mathematical Society
DATE FOUNDED: 1975 FREQUENCY: Quarterly
PRICE: $10.90 LANGUAGE: English, French
INDEXED/ABSTRACTED: Math R
TARGET READER: Applied Mathematicians and Scientists using mathematics.

Publishes expository papers bridging the gap between professional mathematicians and the users of mathematics.

APPLIED NUMERICAL MATHEMATICS
ISSN: 0168-9274

EDITOR: R. Vichnevetsky
Department of Computer Science
Rutgers University
New Brunswick, NJ 08903
MS REQUIREMENT: Dbl spaced, 10-20 lines of summary
PUBLISHER: North-Holland
DATE FOUNDED: 1985 FREQUENCY: Monthly
PRICE: $475.75 LANGUAGE: English
INDEXED/ABSTRACTED: Appl Mech Rev, Compumath, Math R
TARGET READER: Applied Mathematicians, Computational
Mathematicians, Computer Scientists, Physicists, Engineers

Devoted to contemporary problems in numerical computing. Every complete volume is composed of both general issues of contributed papers and special issues devoted to particular topics. Also publishes papers describing relevant applications in such fields as physics, fluid dynamics, engineering, and other branches of applied science.

APPLIED STOCHASTIC MODELS AND DATA ANALYSIS
ISSN: 8755-0024

EDITOR: J. Janssen
                          Université Libre de Bruxelles
                          Ecole de Commerce SOLVAY
                          Centre d'Analyse des Données
                          et Processus Stochastiques
                          50, Avenue F-D
                          Roosevelt-cp 194/7, B-1050
                          Brussels, BELGIUM
MS REQUIREMENT: Triplicate, dbl spaced, wide margins,
maximum length of contribution is 10,000 words, maximum 200
word summary, include key words
PUBLISHER: John Wiley
DATE FOUNDED: 1985   FREQUENCY: Quarterly
PRICE: $235   MICROFORM: UMI
LANGUAGE: English
INDEXED/ABSTRACTED: Curr Cont, Curr Ind Stat, Eng Ind,
Math R, Zent Math
TARGET READER: Statisticians, Numerical Analysts,
Mathematicians, Operations Researchers, Biometricians, Economists,
Computer Scientists, Data Analysts

Aims to serve as the interface between the theoretical aspects of applied probability and data analysis and their applications in the real world.

ARCHIVE FOR MATHEMATICAL LOGIC
ISSN: 0933-5846

HISTORY: *Archiv für Mathematische Logik und Grundlagenforschung* [ISSN: 0003-9268] Bd1-Bd26 (1950-1987)

EDITOR: H. D. Ebbinghaus
Mathematisches Institut
Abteilung fur Mathematische Logik
und Grundlagen der Mathematik,
Albertstrasse 23B, D-7800
Freiburg, I. Br., FRG

PUBLISHER: Springer-Verlag
DATE FOUNDED: 1988   FREQUENCY: Bi-Monthly
PRICE: $296.91 MICROFORM: UMI
LANGUAGE: English, French, German, Italian, Latin
INDEXED/ABSTRACTED: Curr Math Publ, Math R, Sci Abstr, Zent Math
TARGET READER: Mathematicians, Logicians, Theoretical Computer Scientists, Philosophers who are interested in the applications of mathematical logic

Publishes research papers and occasionally surveys or expository papers on mathematical logic. Covers other related areas such as theoretical computer science or philosophy, as long as the methods of mathematical logic play a significant role.

ARCHIVE FOR RATIONAL MECHANICS AND ANALYSIS
ISSN: 0003-9527

EDITOR: S. S. Antman,
Department of Math
University of Maryland
College Park, MD 20742

PUBLISHER: Springer-Verlag
DATE FOUNDED: 1957   FREQUENCY: Monthly

PRICE: $1287.42   MICROFORM: UMI
REPRINT: ISI   LANGUAGE: English, French,
                                         German, Italian
INDEXED/ABSTRACTED: Appl Mech Rev, Chem Abstr, Eng Ind,
Ind Sci Rev, Math R, Sci Abstr, Sci Cit Ind, Zent Math
TARGET READER: Applied Mathematicians, Physicists, Analysts

Covers the discipline of mechanics as a deductive, mathematical
science. Promotes pure analysis, particularly in contexts of application
of continuum mechanics, thermodynamics, non-linear phenomena and
dynamic systems.

ARKIV FÖR MATEMATIK
ISSN: 0004-2080

HISTORY: *Arkiv für Matematik, Astronomi och Fysik*
                [ISSN: 0365-4133], Bd1-Bd36 (1903-1949)

EDITOR: L. Arkeryd
                     Chalmers University of Technology
                     Göteborg, SWEDEN
MS REQUIREMENT: Send to Publisher
PUBLISHER: Institut Mittag-Leffler
DATE FOUNDED: 1949   FREQUENCY: Semi-Annually
PRICE: $90   LANGUAGE: English, French,
                                          German
INDEXED/ABSTRACTED: Compumath, Math R, Sci Abstr
TARGET READER: Mathematicians in all fields

Publishes original research papers in all fields of mathematics.

ARS COMBINATORIA
ISSN: 0381-7032

EDITOR: Prof. W. L. Kocay
                     Department of Computer Science

University of Manitoba
Winnipeg, CANADA   R3T 2N2
MS REQUIREMENT: Request from Editor
PUBLISHER: Charles Babbage Research Center
DATE FOUNDED: 1976   FREQUENCY: Semi-Annually
PRICE: $42   LANGUAGE: English
CIRC: 300
INDEXED/ABSTRACTED: Compumath, Int Abstr Oper Res, Math R
TARGET READER: Mathematicians, Combinatorial Analysts

Publishes research papers in the area of combinatorics.

ARTIFICIAL INTELLIGENCE
ISSN: 0004-3702

EDITOR: Daniel G. Bobrow
                    Xerox Corporation
                    Palo Alto Research Center
                    3333 Coyote Hill Road
                    Palo Alto, CA 94304
PUBLISHER: Elsevier
DATE FOUNDED: 1970   FREQUENCY: 18/yr.
PRICE: $960.67 CIRC: 1100
LANGUAGE: English MICROFORM: RPI, UMI
INDEXED/ABSTRACTED: Cam Sci Abstr, ISI Curr Cont, Math R,
Socio Abstr
TARGET READER: Artificial Intelligence Specialists, Computer
Scientists, Cognitive Scientists, Psychologists

Publishes basic and applied papers describing mature work involving
computational accounts of aspects of intelligence. Papers that are
heavily mathematical in content are preceded by a less technical
introductory section.

ASTERISQUE*
ISSN: 0303-1179

PUBLISHER: American Mathematical Society
DATE FOUNDED: 1973   FREQUENCY: Monthly
PRICE: (Inst) $190   LANGUAGE: English, French
    (AMS Member) $114
INDEXED/ABSTRACTED: Math R, Zent Math
TARGET READER: Mathematicians in all fields

Contains long papers of high quality, lecture notes, and seminar or conference proceedings, including proceedings of the Bourbaki Seminar.

ASYMPTOTIC ANALYSIS
ISSN: 0921-7134

EDITOR: L. S. Frank
                      Mathematisch Instituut
                      University of Nijmegen
                      Tolrnooiveld, 6525 ED Nijmegen
                      THE NETHERLANDS
MS REQUIREMENT: Duplicate, dbl spaced
PUBLISHER: North-Holland
DATE FOUNDED: 1988   FREQUENCY: 4/yr.
PRICE: $170.11 LANGUAGE: English
                              French (occasionally)
INDEXED/ABSTRACTED: Math R
TARGET READER: Mathematicians and Practitioners involved with differential equations, integral equations, optimization, and numerical analysis.

Publishes original research papers on novel mathematical results in asymptotic analysis.   It is designed to offer mathematicians a concentrated source of information which they might need in formal analysis of asymptotic problems.

AUSTRALIAN JOURNAL OF STATISTICS
ISSN: 0004-9581

EDITOR: C. A. McGilchrist
                       School of Math
                       University of New South Wales
                       P. O. Box 1, Kensington
                       New South Wales, 2033
                       AUSTRALIA
MS REQUIREMENT: Original and two copies, dbl spaced, brief
summary, key words or phrases
PUBLISHER: Australian Statistical Publishing Assoc.
DATE FOUNDED: 1959   FREQUENCY: 3/Yr.
PRICE: $55   CIRC: 1500
Back Issues Available   LANGUAGE: English
INDEXED/ASTRACTED: Math R, SSCI
TARGET READER: Theoretical and Applied Statisticians

Publishes theoretical and applied articles in the areas: Mathematical
Statistics, econometrics and probability theory; New applications of
established statistical methods; Applications of newly-developed
methods; Case histories of interesting practical applications; Studies of
concepts suitable for statistical measurement; Sources and applications
of Australian statistical data; Surveys of the applications of statistics in
broad fields.

BIOMETRICAL JOURNAL
ISSN: 0323-3847

HISTORY: *Biometrische Zeitschrift* [ISSN: 0006-3452]
               Bd1-Bd18(1959-1977)

EDITOR: Heinz Ahrens (Berlin)
SEND MS TO: Publisher
MS REQUIREMENT: Duplicate, dbl spaced, up to 30 pages, 2
summaries, one must be English
PUBLISHER: Karl-Weïerstrass Institut für
                       Mathematik der AdW
DATE FOUNDED: 1977   FREQUENCY: Bi-Monthly
PRICE: $285.87 LANGUAGE: English, French,
                                           German, Russian

INDEXED/ABSTRACTED: Anim Breed Abstr, Biol Abstr,
Compumath, Curr Cont, Math R
TARGET READER: Mathematicians, Biologists, Statisticians

Publishes original papers, summary reports on the latest developments
on mathematical penetration of biosciences, proceedings and book
reveiws. The scope includes papers on new theoretical aspects of
mathematics and its application to biological sciences in the widest
sense (including biology, medicine, agriculture science, forestry) or on
the application of known mathematical and statistical methods to new
areas. These may be methods of mathematical statistics and approaches
to mathematical-cybernetic model building of biological systems with
due consideration of electronic data processing.

BIOMETRICS
ISSN: 0006-341X

HISTORY: *Biometrics Bulletin* [ISSN: 0099-4987]
                  V1-V2 (1945-1946)

EDITOR: K. Hinkelmann
                            Department of Statistics
                            Virginia Polytechnic Institute
                            Blacksburg, VA 24061-0439
PUBLISHER: Biometric Society
DATE FOUNDED: 1947   FREQUENCY: Quarterly
PRICE: (Inst) $80   MICROFORM: UMI
              (Invd) $45   CIRC: 8700
LANGUAGE: English, French, German
INDEXED/ABSTRACTED: Appl Mech Rev, Biol Abstr, Biol Agric
Ind, Chem Abstr, Math R
TARGET READER: Biologists, Statisticians, Applied
Mathematicians

Promotes the use of mathematical and statistical methods in the various
subject-matter disciplines by describing and exemplifying development
in these methods and their application in a form readily assimilable by

experimenters and those concerned primarily with analysis of data. It includes statistical, authoritative expository or review articles, analytical or methodological papers and papers in actual worked examples of the statistical analysis proposed.

*BIOMETRIKA**
ISSN: 0006-3444

PUBLISHER: Cambridge University Press
DATE FOUNDED: 1901   FREQUENCY: Quarterly
PRICE: $84   REPRINTS: UMI
CIRC: 3,700 MICROFORM: UMI
LANGUAGE: English, French, German, Italian
INDEXED/ABSTRACTED: Biol Abstr, Biol Agric Ind, Excerpta Med, Math R, Phys Abstr TARGET READER: Biologists, Statisticians

Publishes theoretical papers of direct or potential value in statistical application.

*BIT**
ISSN: 0006-3835

HISTORY: *Nordisk Tidskrift for Informationsehandling*
                [ISSN: 0006-3835] Bd1-Bd6 (1961-1966)

PUBLISHER: Allen Press
DATE FOUNDED: 1967   FREQUENCY: Quarterly
PRICE: $125   LANGUAGE: English
INDEXED/ABSTRACTED: Comput Rev, Math R, Zent Math
TARGET READER: Computer Scientists, Numerical Analysts

BIT contains two sections: Computer Science and Numerical Mathematics. As for Computer Science the emphasis is on design and analysis of algorithms, programming languages, computer systems, and computation techniques. As for Numerical Mathematics BIT has concentrated on numerical methods for problems in linear algebra, integration and differential and integral equations.

BRITISH JOURNAL OF MATHEMATICAL AND STATISTICAL PSYCHOLOGY
ISSN: 0007-1102

HISTORY: *British Journal of Psychology: Statistical Section,*
                V1-V5 (1947-1952)
                British Journal of Statistical Psychology
                [ISSN: 0950-561X] V6-V17 (1953-1964)

EDITOR: A. D. Lovie
                        BRS Journal Office
                        13A Church Lane
                        East Finchley
                        London N2 8DX, ENGLAND
MS REQUIREMENT: Four copies, dbl spaced, brief abstract
PUBLISHER: British Psychological Society
DATE FOUNDED: 1965   FREQUENCY: Semi-Annually
PRICE: $140   CIRC: 770
LANGUAGE: English REPRINTS: ISI
MICROFORM: Swets & Zeitlinger BV
INDEXED/ABSTRACTED: Curr Cont, Ind Med, Math R, Psychol
Abstr, SSCI, Socio Abstr
TARGET READER: Mathematicians, Statisticians, Psychologists

Publishes papers relating to any area of psychology which has a greater mathematical or statistical aspect of their agrument. New models for psychological processes, new approaches to existing data, critiques of existing models and improved algorithms for estimating the parameters of a model are also included.

BULLETIN (New Series) OF THE AMERICAN MATHEMATICAL SOCIETY
ISSN: 0273-0979

HISTORY: *Bulletin of the New York Mathematical Society*, V1-V3 (1891-1894)
Bulletin of the American Mathematical Society [ISSN: 0002-9904] 2nd Series, V1-V84 (1894-1978)

EDITOR:	Richard S. Palais
	Department of Math
	Brandeis University
	Waltham, MA 02154-9110
PUBLISHER:	American Mathematical Society
DATE FOUNDED:	1979 FREQUENCY: Quarterly
PRICE:	$187 MICROFORM: UMI
(Inst Memb)	$150 CIRC: 27000
(Indv Memb)	$112 LANGUAGE: English

INDEXED/ABSTRACTED: Compumath, Curr Cont, Ind Sci Rev, Math R, Sci Cit Ind

TARGET READER: Mathematicians in all fields, Researchers of contemporary mathematical research.

Contains research expository articles, book reviews, and research announcements.

BULLETIN OF INFORMATICS AND CYBERNETICS
ISSN: 0286-522X

HISTORY:	*Bulletin of Mathematical Statistics* [ISSN: 0007-4993] V1-V9 (1941-1981)
EDITOR:	T. Kitagawa
	Research Institute of
	Fundamental Information Science
	Kyushu University 33
	Fukuoka 812, JAPAN
MS REQUIREMENT:	Original and one copy, dbl spaced, limit to

100 word abstract

PUBLISHER:	Research Assoc. of Statistical Science
DATE FOUNDED:	1982 FREQUENCY: Annual

PRICE: $78   LANGUAGE: English
CIRC: 600
INDEXED/ABSTRACTED: Math R, Sci Abstr, Zent Math
TARGET READER: Statisticians, Applied Mathematicians,
Computer Scientists, Cyberneticists

Articles cover two subjects: 1) Informatics: Science of knowledge and
information processed mainly by computers, including mathematical
theory of computation and program, language theory, information
systems, computer network pattern understanding, information model
and computer elements. 2) Cybernetics: Science of control and
communication in animal and machine, including mathematical
programming, control theory, game theory, learning theory,
communication theory, data analysis, mathematical statistics, stochastic
process and biomathematics.

BULLETIN OF MATHEMATICAL BIOLOGY
ISSN: 0092-8240

HISTORY: *Bulletin of Mathematical Biophysics*
               [ISSN: 0007-4985] V1-V34 (1939-1972)

EDITOR: Lee A. Segel
                         Weizmann Institute of Science
                         Rehovot, 76 100, ISRAEL
MS REQUIREMENT: Original and three copies, dbl spaced,
abstract, introduction and summary
PUBLISHER: Pergamon Press
DATE FOUNDED: 1973   FREQUENCY: 6/yr.
PRICE: $420   LANGUAGE: English
CIRC: 1250   MICROFORM: Pergamon Press
INDEXED/ABSTRACTED: Appl Mech Rev, Biol Abstr, Chem
Abstr, Excerpta Med, Math R, Sci Abstr
TARGET READER: Biologists, Applied Mathematicians

Devoted to the publication of research at the interface of theoretical and
experimental biology. In addition to original research papers and survey

articles, it publishes book and software reviews. A special section, The Forum, is a vehicle for rapid and informal communications between experimental and theoretical biologists.

BULLETIN OF THE AUSTRALIAN MATHEMATICAL SOCIETY*
ISSN: 0004-9727

PUBLISHER: Australian Mathematical Society
DATE FOUNDED: 1969   FREQUENCY: Bi-Monthly
PRICE: $187   LANGUAGE: English
Back Issues Available
INDEXED/ABSTRACTED: Appl Mech Rev, Compumath, Math R, Sci Abstr
TARGET READER: Mathematicians in all fields

Publishes original research papers in all branches of mathematics.

BULLETIN OF THE CALCUTTA MATHEMATICAL SOCIETY*
ISSN: 0008-0659

PUBLISHER: Calcutta Mathematical Society
DATE FOUNDED: 1909   FREQUENCY: Bi-Monthly
PRICE: $100   LANGUAGE: English
CIRC: 1,000
INDEXED/ABSTRACTED: Appl Mech Rev, Math R, Sci Abstr
TARGET READER: Mathematicians in all fields

Publishes original papers in all branches of mathematics.

BULLETIN OF THE LONDON MATHEMATICAL SOCIETY*
ISSN: 0024-6093

PUBLISHER: London Mathematical Society
DATE FOUNDED: 1969   FREQUENCY: Bi-Monthly
PRICE: $210   LANGUAGE: English
CIRC: 1,850   MICROFORM: UMI
INDEXED/ABSTRACTED: Compumath, Curr Cont, Math R, Sci
Cit Ind, Zent Math
TARGET READER: Mathematicians in all fields

Publishes short research papers on a wide variety of topics with the
main emphasis on pure mathematics.

*CALCUTTA STATISTICAL ASSOCIATION BULLETIN**
ISSN: 0008-0683

PUBLISHER: Calcutta Statistical Association
DATE FOUNDED: 1947   FREQUENCY: Quarterly
PRICE: $30   LANGUAGE: English
CIRC: 400
INDEXED/ABSTRACTED: Math R, Stat Theory Meth Abstr
TARGET READER: Statisticians

Publishes research papers in all areas of statistics.

CANADIAN JOURNAL OF MATHEMATICS
ISSN: 0008-414X

EDITOR: D. A. Dawson
                             605 Dunton Tower
                             Carleton University
                             Ottawa, Ontario, CANADA, K1S 5B6
MS REQUIREMENT: Triplicate, dbl spaced, paper should be at
least 15 pages, abstract not to exceed 200 words, 1980 Mathematics
Subject Classification Numbers (1985 Rev.)
PUBLISHER: University of Toronto Press
DATE FOUNDED: 1949   FREQUENCY: Bi-Monthly
PRICE: (Inst) $280   LANGUAGE: English, French

(Indv) $70   MICROFORM: UMI
PUBLICATION CHARGES: $50/Page
INDEXED/ABSTRACTED: Math R
TARGET READER: Mathematicians in all fields

Official publication of the Canadian Mathematical Society. Devoted
entirely to research in mathematics. Publishes original papers at least
15 typed pages in length.

*CANADIAN JOURNAL OF STATISTICS**
ISSN: 0319-5724

PUBLISHER: Statistical Society of Canada
DATE FOUNDED: 1973   FREQUENCY: Quarterly
PRICE: $72.72 LANGUAGE: English, French
CIRC: 1,250
Back Issues Available
INDEXED/ABSTRACTED: Compumath, Curr Ind Stat, Math R,
Stat Theory Meth Abstr
TARGET READER: Mathematicians interested in statistics

Publishes papers on all aspects of statistical sciences.

CANADIAN MATHEMATICAL BULLETIN
ISSN: 0008-4395

EDITOR: Stanley O. Kochman
                        Department of Math and Statistics
                        York University
                        North York, Ontario,
                        CANADA, M3J 1P3
MS REQUIREMENT: Triplicate, dbl spaced, abstract not to
exceed 200 words, 1980 Mathematics Subject Classification Numbers
(1985 Rev.)
PUBLISHER: University of Toronto Press
DATE FOUNDED: 1958   FREQUENCY: Quarterly
PRICE: (Inst) $140   LANGUAGE: English,
French (Memb) $35   PUB. CHARGES: $50/page

CIRC: 1,000   MICROFORM: UMI
INDEXED/ABSTRACTED: Math R
TARGET READER: Mathematicians in all fields

Publishes shorter (less than 15 pages) original, research papers in mathematics.

*CARIBBEAN JOURNAL OF MATHEMATICS**
ISSN: 0253-3405

PUBLISHER: University of West Indies
DATE FOUNDED: 1982   FREQUENCY: Semi-Annually
PRICE: $30   LANGUAGE: English
INDEXED/ABSTRACTED: Math R
TARGET READER: Mathematicians

Publishes mathematical papers in all areas.

CHINESE JOURNAL OF MATHEMATICS
ISSN: 0379-7570

EDITORIAL   Chinese Journal of Math
OFFICE: The Mathematical Research
                        Promotion Center
                        Taipei, Taiwan 10764
                        REPUBLIC OF CHINA
MS REQUIREMENT: Duplicate, brief abstract (maximum 200
words)
PUBLISHER: National Taiwan University
DATE FOUNDED: 1973   FREQUENCY: Quarterly
PRICE: $30   LANGUAGE: English, French
INDEXED/ABSTRACTED: Math R, Stat Theory Meth Abstr
TARGET READER: Mathematicians in all fields.

Publishes mathematical research papers in all areas.

CHINESE JOURNAL OF NUMERICAL MATHEMATICS AND APPLICATIONS
ISSN: 0899-4358

EDITOR: Feng Kang
                    Computer Center
                    Academia Sinica
                    Beijing, CHINA
PUBLISHER: Allerton Press
DATE FOUNDED: 1988   FREQUENCY: Quarterly
PRICE: $290   LANGUAGE: English
INDEXED/ABSTRACTED: Math R
TARGET READER: Researchers who are interested in linear and non-linear algebra, numerical analysis.

Covers mathematical research in China, including numerical linear and non-linear algebra and analysis.

COLLEGE MATHEMATICS JOURNAL
ISSN: 0746-8342

HISTORY: *Two-Year College Mathematics Journal*
                [ISSN: 0049-4925] V1-V14 (1970-1983)

EDITOR: Ann Watkins
                    California State University
                    Northridge, CA 91330
MS REQUIREMENT: Five copies, dbl spaced
PUBLISHER: Mathematical Association of America
DATE FOUNDED: 1984   FREQUENCY: Bi-Monthly
PRICE: $85   CIRC: 10,000
LANGUAGE: English MICROFORM: UMI
REPRINTS: UMI
INDEXED/ABSTRACTED: CIJE, Educ Ind, Gen Sci Ind, Math R
TARGET READER: High School Teachers, Undergraduate Mathematics Students

Publishes lively, well-motivated papers that can enrich undergraduate instruction and enhance classroom learning. Articles on mathematics, curriculum and pedagogy, problems and solutions, classroom notes, and a special section on computers, focusing on the earlier years of college-level mathematics.

COLLOQUIUM MATHEMATICUM
ISSN: 0010-1354

EDITORIAL	Colloquium Mathematicum
OFFICE:	Pl. Grunwaldzki 2/4
	50-384 Wroclaw, POLAND
MS REQUIREMENT:	Original and one copy
PUBLISHER:	Polish Academy of Sciences
	Institute of Mathematics
DATE FOUNDED:	1947 FREQUENCY: Irr
PRICE:	Varies CIRC: 900
LANGUAGE:	English, French, German, Russian

INDEXED/ABSTRACTED: Math R
TARGET READER: Mathematicians in all fields

Publishes research papers in all fields of mathematics and its applications including communication on new results, new proofs of known theorems, survey articles, historical notes, programs of research, and open problems.

*COMBINATORICA**
ISSN: 0209-9683

PUBLISHER:	Janos Bolyai Mathematical Society
DISTRIBUTED BY:	Springer-Verlag
DATE FOUNDED:	1981 FREQUENCY: Quarterly
PRICE:	$153.08 LANGUAGE: English

INDEXED/ABSTRACTED: Comput Control Abstr, Elect Electron Abstr, Math R, Phys Abstr
TARGET READER: Combinatorial Analysts, Algebraists

Publishes papers from all branches of combinatorics, as well as from adjacent areas of mathematics.

COMMENTARII MATHEMATICI HELVETICI*
ISSN: 0010-2571

PUBLISHER: Swiss Mathematical Society
DISTRIBUTED BY: Springer-Verlag
DATE FOUNDED: 1929 FREQUENCY: Quarterly
PRICE: $204.20 LANGUAGE: English, French, German
INDEXED/ABSTRACTED: Compumath, Ind Sci Rev, Math R, Sci Cit Ind
TARGET READER: Mathematicians in all areas

Publishes original papers on all aspects of pure and applied mathematics.

COMMENTARII MATHEMATICI UNIVERSITATIS SANCTI PAULI*
ISSN: 0010-258X

DISTRIBUTED BY: Kinokuniya Co., Ltd
DATE FOUNDED: 1952 FREQUENCY: Semi-Annually
PRICE: $182 LANGUAGE: English
INDEXED/ABSTRACTED: Math R
TARGET READER: Mathematicians in all areas

Publishes original papers in all aspects of mathematics.

COMMUNICATIONS IN ALGEBRA
ISSN: 0092-7872

EDITOR: Earl J. Taft
Rutgers University
New Brunswick, NJ 08903

MS REQUIREMENT   Original and two copies, dbl spaced
PUBLISHER: Marcel Dekker
DATE FOUNDED: 1974   FREQUENCY: Monthly
PRICE: (Inst) $985   MICROFORM: RPI
              (Indv) $492.50 LANGUAGE: English
INDEXED/ABSTRACTED: Compumath, Curr Cont, Ind Sci Rev,
Math R, Sci Cit Ind
TARGET READER: Algebraists

Presents full-length articles that reflect significant advances in all areas
of current algebraic interest and activity with the exception of topics
dealing exclusively with classical number theory.

COMMUNICATIONS IN APPLIED NUMERICAL METHODS
ISSN: 0748-8025

EDITORS: Roland W. Lewis
                      Department of Civil Engineering
                      University of Wales
                      Swansea, SA2 8PP ENGLAND
        and   Graham F. Carey
                      College of Engineering/WRW305
                      University of Texas
                      Austin, TX 78712
MS REQUIREMENT: Triplicate, dbl spaced, maximum length of
paper is 2500 words, 200 word summary
PUBLISHER: John Wiley
DATE FOUNDED: 1985   FREQUENCY: 8/Yr.
PRICE: $235   LANGUAGE: English
MICROFORM: UMI
INDEXED/ABSTRACTED: Cam Sci Abstr, Compumath, Curr Cont,
Math R, Zent Math
TARGET READER: Practicing Engineer Researchers and Educators
in numerical and computational methods

Publishes short, referred contributions describing significant
developments in numerical methods and the application of such

technique to the solution of practical engineering problems. Another category of the journal will contain contributions on the industrial application of numerical methods. Longer articles will be published in its companion journal *International Journal for Numerical Methods in Engineering*.

COMMUNICATIONS IN MATHEMATICAL PHYSICS
ISSN: 0010-3616

CHIEF EDITOR:	A. Jaffe
	Lyman Laboratory of Physics
	Harvard University
	Cambridge, MA 02138
MS REQUIREMENT:	Duplicate, dbl spaced, abstract
PUBLISHER:	Springer-Verlag

DATE FOUNDED:	1966	FREQUENCY:	Semi-Monthly
PRICE:	$2887	MICROFORM:	UMI
REPRINT:	ISI	LANGUAGE:	English, French, German

INDEXED/ABSTRACTED: Comput Control Abstr, Elect Electron Abstr, Math R, Phys Abstr
TARGET READER: Mathematical Physicists, Mathematicians interested in Lie group

Publishes high level, theoretical mathematical physics papers covering a broad spectrum from classical to quantum physics.

COMMUNICATIONS IN PARTIAL DIFFERENTIAL EQUATIONS
ISSN: 0360-5302

EDITOR:	M. G. Crandall
	Department of Math
	University of California
	Santa Barbara, CA 93106
MS REQUIREMENT:	Original and two copies
PUBLISHER:	Marcel Dekker

DATE FOUNDED: 1976   FREQUENCY: Monthly
PRICE: $575   MICROFORM: RPI
LANGUAGE: English, French, German, Russian
INDEXED/ABSTRACTED: Compumath, Math R, Ref Zh, Zent Math
TARGET READER: Applied Mathematicians, Differential Equation Theorists

Offers collections of articles in the mathematical aspects of partial differential equation and applications, including the theory of linear and non-linear equations.

COMMUNICATIONS IN STATISTICS: SIMULATION AND COMPUTATION.
ISSN: 0361-0918

HISTORY: *Communications in Statistics*
             [ISSN: 0090-3272] V1-V4 (1973-1975)

EDITOR: D. B. Owen
                  Department of Statistical Science
                  Southern Methodist University
                  Dallas, TX 75275-0332
PUBLISHER: Marcel Dekker
DATE FOUNDED: 1976   FREQUENCY: Quarterly
PRICE: (Inst) $365   LANGUAGE: English
           (Indv) $182   MICROFILM: RPI
INDEXED/ABSTRACTED: Compumath, Ind Sci Rev, Math R, Sci Abstr, Sci Cit Ind
TARGET READER: Statisticians, Computer Scientists

Deals with problems at the interface of statistics and computer science. Devoted to presenting the formulation and discussion of problems as well as their solutions.

COMMUNICATIONS IN STATISTICS:
STOCHASTIC MODELS

ISSN: 0882-0287

EDITOR:	Marcel F. Neuts
	Department of Systems
	and Industrial Engineering
	University of Arizona
	Tucson, AZ 85721

MS REQUIREMENT: Original and two copies, 1½ line spaced, brief abstract.

PUBLISHER:		Marcel Dekker		
DATE FOUNDED:		1985	FREQUENCY:	Quarterly
PRICE:	(Inst)	$255	LANGUAGE:	English
	(Indv)	$127.50	MICROFORM:	RPI

INDEXED/ABSTRACTED: Curr Ind Stat, Curr Cont, J Cont Quan Meth, Math R, Stat Theory Meth Abstr

TARGET READER: Probabilists, Natural Scientists

Publishes original papers devoted to the theory and the applications of probability as they arise in the modelling of phenomena in the Natural Sciences and Technology.

COMMUNICATIONS IN STATISTICS: THEORY AND METHODS
ISSN: 0361-0926

HISTORY:	*Communications in Statistics*
	[ISSN: 0090-3272] V1-V4 (1973-1975)
EDITOR:	D. B. Owen
	Department of Statistical Science
	Southern Methodist University
	Dallas, TX 75275-0332

PUBLISHER:		Marcel Dekker		
DATE FOUNDED:		1976	FREQUENCY:	Monthly
PRICE:	(Inst)	$985	Language:	English
	(Indv)	$492.50	MICROFORM:	RPI

INDEXED/ABSTRACTED: Compumath, Ind Sci Rev, Math R, Sci Abstr, Sci Cit Ind

TARGET READER: Statisticians and Computer Scientists

New applications of known statistical methods to actual problems in
industry and government as well as a math orientation to statistical
studies.

COMMUNICATIONS ON PURE AND APPLIED MATHEMATICS
ISSN: 0010-3640

EDITOR:	Natascha A. Brunswick
	Courant Institute of
	Mathematical Sciences
	New York University
	New York, NY 10012
PUBLISHER:	John Wiley

DATE FOUNDED: 1939   FREQUENCY: 9/yr
PRICE: $435   LANGUAGE: English
MICROFORM: RPI   CIRC: 1400
INDEXED/ABSTRACTED: Appl Mech Rev, Compumath, Curr
Cont, Math R, SCI,
TARGET READER: Pure and Applied Mathematicians, Physicists,
Computer Scientists, and Statisticians.

Devoted to mathematical contributions to the sciences, both theoretical
and applied papers of original or expository type. Papers originate at
or are solicited by the Courant Institute of Mathematical Sciences.

COMPLEX VARIABLES THEORY AND APPLICATION*
ISSN: 0278-1077

PUBLISHER: Gordon and Breach
DATE FOUNDED: 1982   FREQUENCY: Irr
PRICE: Request   LANGUAGE: English
            from Publisher MICROFORM: Publisher
INDEXED/ABSTRACTED: Math R

TARGET READER: Complex Analysts

Publishes papers in the areas of complex analysis.

*COMPUMATH CITATION INDEX**
ISSN: 0730-6199

PUBLISHER: Ínstitute for Scientific Information
DATE FOUNDED: 1981   FREQUENCY: 3/Yr.
PRICE: $1495   LANGUAGE: English
MICROFORM: Available in magnetic tape and online
INDEXED/ABSTRACTED: Math R
TARGET READER: Mathematicians, Computer Scientists,
Statisticians

A multidisciplinary index to the journal literature of computer science, mathematics, statistics, and operations research, and other related disciplines such as mathematical physics and econometrics.

COMPUTATIONAL COMPLEXITY
ISSN: 1016-3328

EDITOR: Joachim von zur Gathen
                                  Department of Computer Science
                                  University of Toronto
                                  Toronto, Ontario M5S 1A4, CANADA
                                  Tel: (416) 978-6024
                                  FAX: (416) 978-4765
MS REQUIREMENT: Four copies, dbl spaced, abstract less than
150 words, 4 to 6 key words, encourage camera ready copy, accept
LATEX or TEX
PUBLISHER: Birkhaüser-Verlag
DATE FOUNDED: 1991   FREQUENCY: 4/yr.
PRICE: $198   LANGUAGE: English
TARGET READER: Theoretical Computer Scientists

Presents outstanding research in computational complexity. Its subject is at the interface between mathematics and theoretical computer science, with a clear mathematical profile and a strict mathematical format.

COMPUTATIONAL GEOMETRY: THEORY AND APPLICATIONS
ISSN: 0925-7721

EDITOR: Jörg Rüdiger Sack
                            School of Computer Science
                            Carleton University
                            Ottawa, Ontario, CANADA K1S 5B6
PUBLISHER: North Holland
DATE FOUNDED: 1991   FREQUENCY: Bi-Monthly
PRICE: $157.30 LANGUAGE: English
TARGET READER: Computer Scientists, Geometricians

Publishes research papers in theoretical and applied aspects of computational geometry. It publishes and disseminates information on the applications, techniques, and use of computational geometry.

COMPUTATIONAL MATHEMATICS AND MODELING
ISSN: 1046-283X

EDITOR: W. A. Light
                            Department of Math
                            University of Lancaster
                            Lancaster, LA1 4YL, ENGLAND
TRANSLATED BY: Zvi Lerman
PUBLISHER: Consultants Bureau
DATE FOUNDED: 1990   FREQUENCY: Quarterly
PRICE: $295   LANGUAGE: English
INDEXED/ABSTRACTED: Comput Control Abstr, Elect Electron Abstr

TARGET READER: Computer Scientists, Numerical Analysts

Translation of selected articles from *Sbornik Trudov Fakul'teta Vychislitel'noi Matematiki i Kibernetiki Moskovskogo Gosudarstvennogo Universiteta* (Translation of the faculty of Computational Mathematics and Cybernetics of the Moscow State University.) From time to time, translations of significant articles from other Russian sources will also be included.   Covers subjects of discrete mathematics, numerical analysis and computational number theory.

COMPUTATIONAL STATISTICS AND DATA ANALYSIS
ISSN: 0167-9473

EDITOR: Stanley P. Azen
                         Department of Preventive Medicine
                         University of Southern California
                         1420 San Pablo St., PMB B1010
                         Los Angeles, CA 90033
MS REQUIREMENT: Four copies, dbl spaced, 200-word abstract, list of key words, three copies for informational articles and news items
PUBLISHER: North-Holland
DATE FOUNDED: 1983   FREQUENCY: Bi-Monthly
PRICE: $384.26 CIRC: 1000
LANGUAGE: English
INDEXED/ABSTRACTED: ACM Guide Comput Lit, Curr Cont Stat, Eng Ind, Math R, Sci Abstr
TARGET READER: Statisticians, Software Managers, Researchers in social, engineering, natural, physical, and medical sciences, Survey and Market Researchers

Dedicated to the dissemination of methodological research and applications in the areas of computational statistics and data analysis. It consists of three refereed sections and a section of news on statistical computing. The refereed sections are: 1) Data Analytic Methodology and Procedures; 2) Applications and Comparative Studies; 3) Computational Statistics.

COMPUTERS AND MATHEMATICS WITH APPLICATION

ISSN: 0898-1221

HISTORY: *Computers and Mathematics with Application*
                [ISSN: 0097-4943) V1-V11 (1975-1985)
                Computer and Mathematics with Applications.
                Part A [ISSN: 0886-9553] V12 (1986)
                Computer and Mathematics with Applications.
                Part B [ISSN: 0886-9561] V12 (1986)

EDITOR: Ervin Y. Rodin
                     Department of Systems Science
                     and Mathematics,
                     P. O. Box 1040
                     Washington University
                     St. Louis, MO 63130
MS REQUIREMENT: Dbl spaced, short abstract, will accept
camera ready papers
PUBLISHER: Pergamon Press
DATE FOUNDED: 1987   FREQUENCY: Bi-Weekly
PRICE: $1120   CIRC: 1125
LANGUAGE: English MICROFORM: MIM, UMI
INDEXED/ABTRACTED: Appl Mech Rev, Compumath, Curr Cont,
Math R, Sci Abstr, Sci Cit Ind
TARGET READER: Applied Mathematicians, Computer Scientists,
Programmers, System Scientists.

Provides a medium of exchange for fields where a non-trivial interplay
between mathematics and computers exists. Three principal areas of
interest are: 1) Computers in mathematical research; 2) Mathematical
models of computer systems; 3) Interactive applications. The journal
attempts to bring together original results, but theoretical and applied as
well as occasional tutorial, review, or educational articles.

CONSTRUCTIVE APPROXIMATION

ISSN: 0176-4276

EDITOR: E. B. Saff
                           Department of Math
                           University of So. Florida
                           Tampa, FL 33620
MS REQUIREMENT: Triplicate, brief abstract, key words and
phrases, current AMS MOS Classification Numbers
PUBLISHER: Springer-Verlag
DATE FOUNDED: 1985   FREQUENCY: Quarterly
PRICE: $138   MICROFORM: UMI
LANGUAGE: English REPRINT: ISI
INDEXED/ABSTRACTED: Compumath, Curr Cont, Math R, Zent
Math
TARGET READER: Theoretical Mathematicians doing research in
function theory, functional analysis, harmonic analysis, and other
fields which interface with appoximation, Engineers, Computer
Scientists

Devoted to significant research in approximations and expansions. It
emphasizes those techniques of approximation theory that are most
useful in either numerical computation or in the constructive aspects of
mathematical analysis.

*CRYPTOLOGIA**
ISSN: 0161-1194

PUBLISHER: Rose-Hulman Institute of Technology
DATE FOUNDED: 1977   FREQUENCY: Quarterly
PRICE: $34   LANGUAGE: English
MICROFORM: UMI   REPRINTS: UMI
CIRC: 1,000
INDEXED/ABSTRACTED: Comput Control Abstr, Elect Electron
Abstr, Math R, Phys Abstr
TARGET READER: Mathematicians interested in cryptology,
Computer Scientists

Publishes research papers on all aspects of cryptology including
computer security, mathematics, codes, cryptoanalysis and ancient
languages.

CURRENT MATHEMATICAL PUBLICATIONS
ISSN: 0361-4794

HISTORY: *New Publications - American Mathematical*
Society [ISSN: 0002-9912) V1-V32
(1964-1971) merged with
Contents of Contemporary Mathematical
Journals [ISSN: 0010-759X] V1-V3 (1969-1971)
Contents of Contemporary Mathematical Journals
and New Publications, V4-V6 (1972-1974)

EDITORIAL	Mathematical Reviews
OFFICE:	416 Fourth Street/P. O. Box 8604
	Ann Arbor, MI 48107-9-8604
	Tel: (313) 996-5250
	FAX: (313) 996-2916
PUBLISHER:	American Mathematical Society
DATE FOUNDED:	1975 FREQUENCY: 17/yr.
PRICE: (Inst)	$314 LANGUAGE: English
(Indv)	$188
(Inst Memb)	$251
(Reviewer)	$126

INDEXED/ABSTRACTED: Math R
TARGET READER: Mathematicians in all fields

Contains a subject classified index of papers and books being published currently in mathematics. The subject listing includes names of authors and titles of items, classified by the Mathematical Reviews' editors according to the 1991 Mathematics Subject Classification.   Author Index and key index are included in each issue.

Math Sci is the electronic online version of CMP and MR. MathDoc is the fee-based document delivery service designed for users of CMP and MR.

CZECHOSLOVAK MATHEMATICAL JOURNAL
ISSN: 0011-4642

HISTORY: *Casopis Pestovani Matematiky a Fysiky*
Casopis Pro Pestovani Matematiky,
V1-V2 (1951-1952)
Chekoslovaskii Matematicheskii Zhurnal
[ISSN: 0528-9181] V1-V18 (1951-1968)

EDITOR: Miroslav Fiedler
Mathematical Institute
of the Czechoslovak Academy of Sciences
Zitna 25, 115 67, Praha 1,
CZECHOSLOVAKIA

PUBLISHER: Mathematical Institute of the
Czechoslovak Academy of Sciences
Distributed by Plenum Publishing Corp.

DATE FOUNDED: 1969 FREQUENCY: Quarterly
PRICE: $455 CIRC: 1200
LANGUAGE: English, French, German, Russian
INDEXED/ABSTRACTED: Appl Mech Rev, Compumath, Comput Rev, Math R
TARGET READER: Mathematicians in all fields.

Publishes original mathematical research papers, news, and notices.

DIFFERENTIAL AND INTEGRAL EQUATIONS
ISSN: 0893-4983

EDITOR: Reza Aftabizadeh
Department of Math
Ohio University
Athens, OH 45701

MS REQUIREMENT: Duplicate, up to 150 word abstract, AMS Subject Classification Numbers.
PUBLISHER: Ohio University Press
DATE FOUNDED: 1988 FREQUENCY: Bi-Monthly
PRICE: $260 LANGUAGE: English
Back Issues are Available
INDEXED/ABSTRACTED: Math R
TARGET READER: Mathematicians involved with differential

equations and integral equations.

Publishes research papers in the areas of ordinary differential equations, functional differential equations, partial differential equations, integral equations, and applications.

DIFFERENTIAL EQUATIONS
ISSN: 0012-2661

Translation of *Differentsial' nye Uravneniya*

PUBLISHER: Consultants Bureau
DATE FOUNDED: 1965   FREQUENCY: Monthly
PRICE: $995
INDEXED/ABSTRACTED: Appl Mech Rev, Compumath, Comput Inf Syst Abstr J, Curr Cont, Math R, Zent Math
TARGET READER: Mathematicians interested in differential equations.

A publication of the Academy of Sciences of the USSR. Devoted exclusively to Soviet research in the field of differential equations. Contributions focus on theoretical and applied aspects of ordinary differential equations and partial differential equations.

DIFFERENTIAL GEOMETRY AND ITS APPLICATIONS*
ISSN: 0926-2245

EDITOR: D. Krupka, Department of Math
                     Janackovo nam 2a
                     66295 Brno, CZECHOSLOVAKIA
PUBLISHER: North-Holland
DATE FOUNDED: 1991   FREQUENCY: 4/Yr.
PRICE: $135   LANGUAGE: English
TARGET READER: Mathematicians who are interested in global differential geometry, differential equations on manifolds

Publishes original research papers and survey papers in differential
geometry and in all interdisciplinary areas in mathematics which use
differential geometric methods and investigate geometrical structures.

DISCRETE AND COMPUTATIONAL GEOMETRY
ISSN: 0179-5376

EDITOR: Jacob E. Goodman
                  Department of Math
                  City College, CUNY
                  New York, NY 10031
                  Tel: (212) 650-5141
PUBLISHER: Springer-Verlag
DATE FOUNDED: 1986   FREQUENCY: Bi-Monthly
PRICE: $171   MICROFORM: UMI
LANGUAGE: English
INDEXED/ABSTRACTED: Compumath, Curr Cont, Eng Ind, Math
R, Zent Math
TARGET READER: Mathematicians in the area of geometry and
combinatorics, Computer Scientists

An international journal of mathematics and computer science covering
a broad range of topics in which geometry plays a fundamental role.
Publishes research articles in the areas of combinatorial geometry,
design and analysis of geometric algorithms, convex polytopes,
multi-dimensional searching and sorting, extremal geometric problems,
etc.

DISCRETE APPLIED MATHEMATICS
ISSN: 0166-218X

HISTORY: *Discrete Applied Mathematics,*
                 V1-V10 (1979-1985)
               Discrete Applied Mathematics and Combinatorial
                 Operations Research, V11-V20 (1985-1988)

EDITOR: Peter L. Hammer
                               Centre for Operations Research
                               Hill Center for the Mathematical
                               Sciences/Busch Campus
                               Rutgers University
                               New Brunswick, NJ 08903
SEND MS TO: Nelly Segal at above address.
MS REQUIREMENT: Triplicate, dbl spaced, English abstract and
extra French abstract for French papers.
PUBLISHER: North-Holland
DATE FOUNDED: 1988   FREQUENCY: 15/yr.
PRICE: $756.70 MICROFILM: UMI, RPI
LANGUAGE: English, French
INDEXED/ABSTRACTED: Compumath, Curr Cont, Cyb Abstr,
Eng Ind, Math R, Sci Abstr, Sci Cit Ind, Zent Math
TARGET READER: Researcher in discrete optimization, game
theory, etc. Biologists, Sociologists, Chemists, and Psychologists
who use mathematics in their work, Computer Scientists

Brings together research papers in different areas of discrete applicable
mathematics and applications of it. Contributions can be research
papers, short notes, surveys, and possibly research problems.

DISCRETE MATHEMATICS
ISSN: 0012-365X

EDITOR: Peter L. Hammer
                               Centre for Operations Research
                               Hill Center for the Mathematical
                               Sciences/Busch Campus
                               Rutgers University
                               New Brunswick, NJ 08903
SEND MS TO: Nelly Segal at above address.
MS REQUIREMENT: Triplicate, dbl spaced, English abstract and
extra French abstract for French papers.
PUBLISHER: North-Holland
DATE FOUNDED: 1971   FREQUENCY: 30/yr.
PRICE: $1573.97 MICROFORM: RPI, UMI

CIRC: 2000 LANGUAGE: English, French
INDEXED/ABSTRACTED: ACM Guide Comput Lit, Cam Sci
Abstr, ISI Curr Cont, Math R
TARGET READER: Researchers in various fields of discrete
mathematics, such as graph theory, network theory, coding theory, or
combinatorics.

Publishes research papers, short notes, communications, occasional
survey articles and research problems on different areas of discrete
mathematics.

DISCRETE MATHEMATICS AND APPLICATIONS
ISSN: 0924-9265

EDITOR: V Ya. Kozlov
PUBLISHER: VSP
DATE FOUNDED: 1991 FREQUENCY: 4/yr.
PRICE: (Inst) $423 (Bi-monthly for 1992)
LANGUAGE: English
TARGET READER: Researchers in various fields of discrete
mathematics and computer science

Provides the latest information on the development of discrete
mathematics in the USSR to a world-wide readership. Covers various
subjects in the field such as combinatorial analysis, graph theory,
functional systems theory, coding, probability problems of discrete
mathematics, etc.

DUKE MATHEMATICAL JOURNAL
ISSN: 0012-7094

EDITOR: Morris Weisfeld
SEND MS TO: Publisher
MS REQUIREMENT: Accepts papers in TEX which can be
submitted by electronic mail, 2 copies, dbl spaced.

PUBLISHER: Duke University Press
DATE FOUNDED: 1935   FREQUENCY: 9/yr.
PRICE: (Inst) $396   MICROFORM: MIM, UMI
            (Indv) $198   REPRINT: ISI, UMI
LANGUAGE: English
INDEXED/ABSTRACTED: Compumath, Curr Cont, Ind Sci Rev,
Math R, Sci Cit Ind
TARGET READER: Mathematicians in all fields.

Publishes original research papers in pure and applied mathematics.

EDUCATIONAL STUDIES IN MATHEMATICS
ISSN: 0013-1954

EDITOR: Willibald Dörfler
                  Institut für Mathematik
                  Universität Klagenfurt
                  Universitätsstrasse 65-67
                  A-9022 Klagenfurt, AUSTRIA
MS REQUIREMENT: Four copies, dbl spaced, English abstract
and an abstract in the language in which the article is written
PUBLISHER: Kluwer Academic Publishers
DATE FOUNDED: 1968   FREQUENCY: Bi-Monthly
PRICE: $201.70 MICROFORM: UMI
LANGUAGE: English, French, German
INDEXED/ABSTRACTED: Curr Ind Edu, Math R, Ref Rh, Zent
Math
TARGET READER: Mathematics Educators

Presents new ideas and developments which are considered to be of
major importance to those working in the field of mathematical
education.   Deals with didactical, methodological, and pedagogical
subjects rather than with specific programs for teaching mathematics.

ELEMENTE DER MATHEMATIK
ISSN: 0013-6018

EDITOR: M. Jeger
                       Unter Geissenstein 8
                       CH-6005 Luzern, SWITZERLAND
PUBLISHER: Birkhaüser-Verlag
DATE FOUNDED: 1946   FREQUENCY: Bi-Monthly
PRICE: $62.30 MICROFORM: UMI
LANGUAGE: English, French, German
INDEXED/ABSTRACTED: Math R
TARGET READER: Mathematicians in all fields, Advanced
Students.

Publishes research papers, survey articles and short notes in all fields
of mathematics with particular emphasis on subjects amenable to an
elementary approach.

*EMPLOYMENT INFORMATION IN THE MATHEMATICAL SCIENCES**
ISSN: 0163-3287

HISTORY: *Mathematical Sciences Employment Register,*
          (1960-1972)
         Employment Information for Mathematicians
         [ISSN: 0147-3018] (1972-1978)

PUBLISHER: American Mathematical Society
DATE FOUNDED: 1978   FREQUENCY: 6/yr.
PRICE: (Inst) $135   LANGUAGE: English
           (Indv) $81
           (Stud) $34
TARGET READER: Anybody seeking a career in the mathematics
field

Provides information on available positions in the mathematical
sciences. It is published jointly by the AMS, MAA, and SIAM.

*ERGODIC THEORY AND DYNAMICAL SYSTEMS**
ISSN: 0143-3857

PUBLISHER: Cambridge University Press
DATE FOUNDED: 1981   FREQUENCY: Quarterly
PRICE: $275   LANGUAGE: English
MICROFILM: UMI   CIRC: 450
INDEXED/ABSTRACTED: Compumath, Math R
TARGET READER: Ergodic Theorists, Probabilists, Algebraists

Publishes papers on the application of ergodic theory to differential
geometry, statistical mechanics, number theory, and operator algebra.

EUROPEAN JOURNAL OF COMBINATORICS
ISSN: 0195-6698

SEND MS TO: European Journal of Combinatorics
                              54 Blvd. Raspail,
                              75006 Paris, FRANCE
                              Tel: Paris (1) 49 54 20 41
MS REQUIREMENT: Triplicate, dbl spaced, brief abstract in
English. PUBLISHER: Academic Press
DATE FOUNDED: 1980   FREQUENCY: Bi-Monthly
PRICE: (Inst) $268   LANGUAGE: English
               (Indv) $98
INDEXED/ABSTRACTED: Compumath, Curr Cont, Math R, Sci
Abstr, Sci Cit Ind
TARGET READER: Mathematicians interested in combinatorics,
Computer Scientists, Engineers

Publishes full length research papers, short notes, research problems,
and other information of interest in pure mathematics specializing in
theories from combinatorial problems. Articles deal with mathematical
structures within combinatorics and establishing direct links between
combinatorics and other branches of mathematics and the theories of
computing.

EUROPEAN JOURNAL OF OPERATIONAL RESEARCH
ISSN: 0377-2217

EDITOR: Alan Mercer
                       Management School
                       Lancaster University
                       Lancaster, LA1 4YX ENGLAND
MS REQUIREMENT: Four copies, key words, abstract between
50-150 words.
PUBLISHER: North-Holland
DATE FOUNDED: 1977   FREQUENCY: 18/yr.
PRICE: $974.15 MICROFORM: RPI, UMI
LANGUAGE: English
INDEXED/ABSTRACTED: Curr Cont, Math R, SSIS, Sci Abstr
TARGET READER: Scientists working in Operations Research,
Mathematicians, Econometricians

Publishes high-quality, original papers that contribute to the practice of
decision making, irrespective of whether their content describes an
application or a theoretical development.

EXPOSITIONES MATHEMATICAE
ISSN: 0723-0869

MANAGING   S. D. Chatterji
EDITOR: Department de Math
                       Ecole Polytechnique
                       Federale de Lausanne
                       Ecublens, CH-1015, Lausanne
PUBLISHER: Bibliographisches Institut & F.A.
DATE FOUNDED: 1982   FREQUENCY: Quarterly
PRICE: $209.51 LANGUAGE: English, French,
                                           German
INDEXED/ABSTRACTED: Math R
TARGET READER: Mathematicians in all fields.

Publishes articles in all branches of mathematics which can be original research papers, surveys, expository essays, historical studies concerning 19th and 20th century mathematics. Shorter articles containing new results, new proofs of known theorems or novel points of view published under the Mathematical Notes section.

FIBONACCI QUARTERLY
ISSN: 0015-0517

EDITOR: Gerald E. Bergum
                               South Dakota State University
                               P. O. Box 2201
                               Brookings, SD 57007-0194
MS REQUIREMENT: Duplicate, dbl spaced
PUBLISHER: Fibonacci Association
DATE FOUNDED: 1963   FREQUENCY: Quarterly
PRICE: (Inst Memb) $70   MICROFORM: UMI
          (Indv Memb) $35   CIRC: 900
LANGUAGE: English
INDEXED/ABSTRACTED: Compumath, Curr Cont, Ind Sci Rev, Math R, Sci Cit Ind
TARGET READER: College Teachers and Students

Serves as a focal point for widespread interest in the Fibonacci and related numbers, especially with respect to new results, research proposals, challenging problems, and innovative proofs of old ideas.

FORUM MATHEMATICUM
ISSN: 0933-7741

SEND MS TO: Forum Mathematicum
                               c/o Otto Gerstner,
                               Mathematisches Institut der
                               Universität, Bismarckstr
                               11/2, D-8520, Erlangen
                               GERMANY
MS REQUIREMENT: Duplicate, dbl spaced, short abstract, 1991 Math Subject Classification Numbers

PUBLISHER: Walter de Gruyter, Inc.
DATE FOUNDED: 1989 FREQUENCY: Bi-Monthly
PRICE: $245 CIRC: 500
LANGUAGE: English
INDEXED/ABSTRACTED: Math R, Zent Math
TARGET READER: Mathematicians in all fields, Physicists

Publishes research articles in all fields of pure and applied mathematics, including mathematical physics. Expository surveys are not included.

FUNCTIONAL ANALYSIS AND ITS APPLICATION
ISSN: 0016-2663

Translation of *Funktsional'nyi Analiz i ego Prilozheniya*

PUBLISHER: Consultants Bureau
DATE FOUNDED: 1967 FREQUENCY: Quarterly
PRICE: $695 LANGUAGE: English
INDEXED/ABSTRACTED: Compumath, Curr Cont, Math R, Zent Math
TARGET READER: Analysts

A publication of the Academy of Sciences of the USSR.

FUNDAMENTA MATHEMATICAE*
ISSN: 0016-2736

EDITORIAL Fundamenta Mathematicae
OFFICE: Sniadeckich 8, P. O. Box 137
00-950 Warszawa, POLAND
Telex: 816112 PANIM PL
PUBLISHER: Institute of Mathematics
Polish Academy of Sciences
DATE FOUNDED: 1920 FREQUENCY: Irr
PRICE: Request from LANGUAGE: English
the publisher

INDEXED/ABSTRACTED: Math R
TARGET READER: Mathematicians in all areas.

Publishes papers devoted to set theory, topology, mathematical logic and foundations, real functions, measure and integration, abstract algebra.

FUZZY SETS AND SYSTEMS
ISSN: 0165-0114

EDITOR: C. V. Negoita
                          Department of Computer Science
                          Hunter College/CUNY
                          New York, NY 10021
MS REQUIREMENT: Triplicate, dbl spaced, up to 200 words of abstract, key words required.
PUBLISHER: North-Holland
DATE FOUNDED: 1978   FREQUENCY: 18/yr.
PRICE: $1001.12 MICROFORM: RPI, UMI
LANGUAGE: English
INDEXED/ABSTRACTED: ACM Guide Comput Lit, Curr Ind Stat, Eng Ind, Math R
TARGET READER: Mathematicians, Statisticians, Probabilists, Systems Engineers.

Publishes theoretical and applied papers in many different areas such as artificial intelligence, decision making, control, engineering, information systems, linguistics, logic, pattern recognition, etc. Aims to improve professional communication between scientists and practitioners who are interested in Fuzzy Sets and Systems.

GAME AND ECONOMIC BEHAVIOR
ISSN: 0899-8256

EDITOR: Ehud Kalai
              MEDS Department
              Northwestern University
              2001 Sheridan Road

Evanston, IL 60208
MS REQUIREMENT: Original and two copies, dbl spaced,
abstract, appropriate Classification Numbers
PUBLISHER: Academic Press
DATE FOUNDED: 1989 FREQUENCY: Quarterly
PRICE: $110 LANGUAGE: English
Back Issues Available
INDEXED/ABSTRACTED: Comput Control Abstr, Elect Electron
Abstr, Phys Abstr
TARGET READER: Mathematicians, Biologists, Computer
Scientists, Psychologists

Publishes original and survey papers dealing with game-theoretic
modeling in the social, biological, and mathematical sciences. Papers
published are mathematically rigorous as well as accessible to readers
in related fields. It is to facilitate cross-fertilization between the theory
and application of game-theoretic reasoning.

GEOMETRICAL AND FUNCTIONAL ANALYSIS
ISSN: 1016-443X

PUBLISHER: Birkaüser-Verlag AG
DATE FOUNDED: 1991 FREQUENCY: Quarterly
PRICE: $138 LANGUAGE: English, French,
German
TARGET READER: Geometricians, Functional Analysts

Covers papers on all major topics of geometry and analysis in their
interactions: elliptic operators of manifolds; global variational calculus,
especially related to symplectic geometry; isoperimetric inequalities,
including combinatorial and probability problems, etc.

GLASGOW MATHEMATICAL JOURNAL
ISSN: 0017-0895

HISTORY: *Proceedings of the Glasgow Mathematical*

Association, V1-V7 (1952-1966)

SEND MS TO: Glasgow Mathematical Journal
                       Department of Math
                       University of Glasgow
                       University Gardens
                       Glasgow G12 8QW, UK
MS REQUIREMENT: Duplicate
PUBLISHER: Oxford University Press
DATE FOUNDED: 1967   FREQUENCY: 3/yr.
PRICE: $105   CIRC: 600
LANGUAGE: English (mainly) Back Issues Available
INDEXED/ABSTRACTED: Compumath, Math R
TARGET READER: Mathematicians in all fields.

Publishes original research papers in any branch of pure or applied mathematics.

GRAPHS AND COMBINATORICS
ISSN: 0911-0119

MANAGING   Jim Akiyama
EDITOR: Department of Math
                       Tokai University
                       Hiratsuka, 259-12 JAPAN
                       Tel (0463)58-1211 (x3224)
MS REQUIREMENT: Triplicate, dbl spaced, with abstract.
PUBLISHER: Springer-Verlag
DATE FOUNDED: 1985   FREQUENCY: 4/yr.
PRICE: $264   MICROFORM: UMI
LANGUAGE: English REPRINT: ISI
INDEXED/ABSTRACTED: Compumath, Math R, Sci Abstr., Zent
Math
TARGET READER: Mathematicians interested in classical and
algebraic graph theories, random graphs, set systems, combinatorial
geometry and Ramsey theories.

Publishes research papers and survey articles concerning combinatorial
mathematics. Also covers short communications, research problems and
announcements.

HISTORIA MATHEMATICA
ISSN: 0315-0860

EDITOR:	Eberhard Knobloch
	Technische Universität
	Berlin, GERMANY
SEND MS TO:	David E. Rowe
	Managing Editor
	Historia Mathematica
	Department of Math
	Pace University
	Pleasantville, NY 10570

MS REQUIREMENT: Four copies, dbl spaced, up to 250 words
of English abstract, 5-7 key words, AMS Classification Numbers,
brief biographical paragraph.

PUBLISHER:	Academic Press		
DATE FOUNDED:	1974	FREQUENCY:	Quarterly
PRICE:	$85	LANGUAGE:	English, French,
			German

INDEXED/ABSTRACTED: Am Hist Life, Math R
TARGET READER: Mathematicians, Librarians, Historians

Publishes articles on the history of all aspects of the mathematical
science in all parts of the world and all periods, including theory and
practice, computer science, statistics, cybernetics, operations research,
mathematical technology, both hardware and software, interrelations
with natural sciences, social sciences, humanities, art, and education,
etc. It includes occasional biographies of mathematicians and historians,
articles on organizations and institutions and the interaction among all
facets of mathematical activity and other aspects of culture and society.

HOKKAIDO MATHEMATICAL JOURNAL
ISSN: 0385-4035

HISTORY: *Journal of the Faculty of Science*, Hokkaido
University, Series I, Mathematics
[ISSN: 0018-3482] V1-V22 (1947-1972)

SEND MS TO:	Managing Editor
	Hokkaido Mathematical Journal
	Department of Math
	Faculty of Science
	Hokkaido Univeristy
	Kita-ku, Sapporo, 060 JAPAN
PUBLISHER:	Kinokuniya, Co., Ltd.

DATE FOUNDED: 1972   FREQUENCY: 3/yr.
PRICE: $230   CIRC: 720
LANGUAGE: English, French, German
Back Issues Available
INDEXED/ABSTRACTED: Math R
TARGET READER: Mathematicians in all fields.

Publishes original mathematical research papers in all fields.

HOUSTON JOURNAL OF MATHEMATICS
ISSN: 0362-1588

MANAGING   Vern I. Paulsen
EDITOR: Department of Math
                University of Houston
                Houston, TX 77204-3476
MS REQUIREMENT: Duplicate, dbl spaced, abstract
PUBLISHER: Houston Journal of Mathematics
DATE FOUNDED: 1975   FREQUENCY: Quarterly
PRICE: $90   CIRC: 525
LANGUAGE: English PUBL CHRGS: Yes (Request
                         for institutional support)
INDEXED/ABSTRACTED: Compumath, Elect Electron Abstr, Math
R, Phys Abstr, Sci Abstr

TARGET READER: Mathematicians in all fields.

Publishes referred research papers of moderate length in all areas of mathematics.

ILLINOIS JOURNAL OF MATHEMATICS
ISSN: 0019-2082

EDITOR: Earl R. Berkson
                               Department of Math
                               273 Altgeld Hall
                               1409 West Green Street
                               University of Illinois
                               Urbana, IL 61801
MS REQUIREMENT: Dbl spaced, 1980 Math Subject
Classification Numbers (1985 Rev.)
PUBLISHER: University of Illinois Press
DATE FOUNDED: 1957   FREQUENCY: Quarterly
PRICE: $90   MICROFORM: MIM, UMI
CIRC: 1100   PUBL CHRG: $40/page
LANGUAGE: English, French, German
INDEXED/ABSTRACTED: Compumath, Curr Cont, Math R, Sci Cit Ind
TARGET READER: Mathematicians in all fields.

Publishes basic research papers in pure and applied mathematics.

IMA JOURNAL OF APPLIED MATHEMATICS
ISSN: 0272-4960

HISTORY: *Journal of the Institute of Mathematics and its Application* [ISSN: 0020-2932] V1-V26 (1965-1980)

EDITOR: R. W. Ogden
                               Department of Math

University of Glasgow
Glasgow G12 8QW   SCOTLAND
MS REQUIREMENT: Original and two copies
PUBLISHER: Oxford University Press
DATE FOUNDED: 1981   FREQUENCY: Bi-Monthly
PRICE: $360   CIRC: 1200
LANGUAGE: English
Back Issues Available
INDEXED/ABSTRACTED: Appl Mech Rev, Compumath, Ind Sci
Rev, Math R, Sci Abstr, Sci Cit Ind
TARGET READER: Applied Mathematicians, Statisticians

Publishes papers in all areas of the application of mathematics,
including analytic and numerical treatments of both physical and
non-physical applied mathematical problems arising in industry. Also
includes papers on new developments of existing mathematical methods,
especially those that have relevance to more than one field of
application and new mathematical methods suggested by particular
applications.

*IMA JOURNAL OF MATHEMATICAL CONTROL & INFORMATION**
ISSN: 0265-0754

EDITOR: J. E. Marshall
                       School of Mathematical Sciences
                       University of Bath
                       ENGLAND
PUBLISHER: Oxford University Press
DATE FOUNDED: 1984   FREQUENCY: Quarterly
PRICE: $185   LANGUAGE: English
INDEXED/ABSTRACTED: Compumath, Comput Control Abstr,
Elect Electron Abstr, Math R, Phys Abstr
TARGET READER: Mathematicians, Computer Scientists

Publishes original papers in mathematical control theory, systems theory
and allied information sciences.

*IMA JOURNAL OF MATHEMATICS APPLIED IN BUSINESS AND INDUSTRY**
ISSN: 0953-0061

HISTORY: *IMA Journal of Mathematics in Management*
            [ISSN: 0268-1129] V1 (1987)

EDITOR: L. C. Thomas
                     Department of Business Studies
                     University of Edinburgh
                     SCOTLAND
PUBLISHER: Oxford University Press
DATE FOUNDED: 1989   FREQUENCY: 4/yr.
PRICE: $160   LANGUAGE: English
TARGET READER: Applied Mathematicians, Econometricians

Publishes relevant and topical applications of mathematical modelling
and mathematical techniques to problems in all areas of business and
industry. Also covers advances in theory which relate to such problems
and will consider case studies based on substantive mathematical
analysis. Research in applied mathematics, operational research,
engineering mathematics, and mathematical economics is included.

*IMA JOURNAL OF MATHEMATICS APPLIED IN MEDICINE AND BIOLOGY**
ISSN: 0265-0746

EDITOR: R. W. Hiorns
                     Department of Statistics
                     University of Oxford
                     ENGLAND
PUBLISHER: Oxford University Press
DATE FOUNDED: 1984   FREQUENCY: Quarterly
PRICE: $185   LANGUAGE: English
INDEXED/ABSTRACTED: Biol Abstr, Compumath, Curr Cont,
Excerpta Med, Math R, Sci Cit Ind

TARGET READER: Mathematicians, Medical and Biological
Researchers

Publishes mathematical papers which can be used in medical and
biological research with emphasis on the special insights.

IMA JOURNAL OF NUMERICAL ANALYSIS
ISSN: 0272-4979

HISTORY: *Journal of the Institute of Mathematics and Its*
           Applications [ISSN: 0020-2932] V1-V26
           (1965-1980)

EDITOR: I. S. Duff, Atlas Centre
                           Rutherford Appleton Laboratory
                           Oxford OX11 0OX, ENGLAND
MS REQUIREMENT: Triplicate, dbl spaced
PUBLISHER: Oxford University Press
DATE FOUNDED: 1981   FREQUENCY: Quarterly
PRICE: $230   LANGUAGE: English
INDEXED/ABSTRACTED: Comput Control Abstr, Elect Electron
Abstr, Math R, Phys Abstr
TARGET READER: Numerical Analysts

Publishes original contributions to all fields of numerical analysis.
Articles will be accepted which treat the theory, development or use of
practical algorithms and the interactions between these aspects.
Occasional survey articles will also be published.

INDAGATIONES MATHEMATICAE
ISSN: 0019-3577

HISTORY: *Koninklijke Akademie van Wetenschappen.*
                     Afdeeling Natuurkundige, V1-V53 (1898-1950)
                   Proceedings. Series A: Mathematical Sciences,
                     V54-V70 (1951-1967)
                   Proceedings of the Koninklijke Nederlandse
                   Akademie van Wetenschappen.

Series A: Mathematical Sciences
[ISSN: 0023-3358] V70#2-V92 (1967-1989)

EDITOR: Editorial Department
                           Koninklÿke Nederlandse
                           Akademie Van Wetenschappen
                           P. O. Box 19121
                           1000 GC Amsterdam
                           THE NETHERLANDS
MS REQUIREMENT: Duplicate, brief summary, abstract up to
100 words
PUBLISHER: North-Holland
DATE FOUNDED: 1990   FREQUENCY: Quarterly
PRICE: $183.70 CIRC: 1500
LANGUAGE: English, French, German
INDEXED/ABSTRACTED: Chem Abstr, Compumath, Curr Cont,
Math R TARGET READER: Mathematicians in all fields.

An official journal for Mathematical Sciences of the Royal
Netherlands Academy of Arts and Sciences.

INDIAN JOURNAL OF PURE AND APPLIED
MATHEMATICS
ISSN: 0019-5588

EDITOR: B. L. S. Prakasa Rao
               Indian Statistical Institute
               7, SJS Sansanwal Marg
               New Delhi   110 016, INDIA
MS REQUIREMENT: Triplicate, dbl spaced, 1980 AMS Subject
Classification Numbers, Maximun 250 word abstract.
PUBLISHER: Indian National Science Academy
DATE FOUNDED: 1970   FREQUENCY: Monthly
PRICE: $200   LANGUAGE: English
INDEXED/ABSTRACTED: Curr Cont, Math R, Sci Cit Ind

TARGET READER: Mathematicians in all fields.

Devoted to original research in pure and applied mathematics, statistics, and related mathematical sciences.

INDIANA UNIVERSITY MATHEMATICS JOURNAL
ISSN: 0022-2518

HISTORY: *Journal of Rational Mechanics and Analysis,*
         V1-V5 (1952-1956)
         Journal of Mathematics and Mechanics,
         V6-V19 (1957-1970)

EDITOR: E. Bedford
                           Department of Math
                           Indiana Univeristy
                           Bloomington, IN 47405
MS REQUIREMENT: Duplicate of manuscript
PUBLISHER: Indiana University Mathematics Journal
DATE FOUNDED: 1970   FREQUENCY: Quarterly
PRICE: $95   LANGUAGE: English, French
PUBLICATION CHANGES: $40/Page (by author's institution)
Back issues available
INDEXED/ABSTRACTED: Compumath, Curr Cont, Eng Ind, Math R, Sci Cit Ind, Zent Math
TARGET READER: Mathematicians of all fields.

Publishes significant research articles in both pure and applied mathematics.

INDUSTRIAL MATHEMATICS
ISSN: 0019-8528

EDITOR: Robert Schmidt
                           Industrial Mathematics
                           P. O. Box 159
                           Roseville, MI 48066

MS REQUIREMENT: Triplicate, dbl spaced, abstract, brief
summary
PUBLISHER: Industrial Mathematics Society
DATE FOUNDED: 1950   FREQUENCY: Semi-Annually
PRICE: $15   CIRC: 600
MICROFORM: UMI   LANGUAGE: English
PUBLICATION CHARGES: $30/Page
                         (Charge Author's Institution)
INDEXED/ABSTRACTED: Appl Mech Rev, Eng Ind, Math R, Sci
Abstr
TARGET READER: Applied Mathematicians

Publishes original papers in the field of applied mathematics containing
novel items of mathematical or scientific interest.

INFORMATION AND COMPUTATION
ISSN: 0890-5401

HISTORY: *Information and Control*
         [0019-9958] V1-V17 (1957-1986)

EDITOR: Albert R. Meyers
                     MIT Laboratory for Computer Science
                     NE43-315
                     545 Technology Square
                     Cambridge, MA 02139
                     Tel: (617) 253-5936
MS REQUIREMENT: Four copies (including original), dbl
spaced, abstract within 150 words
PUBLISHER: Academic Press
DATE FOUNDED: 1987   FREQUENCY: Monthly
PRICE: $510   LANGUAGE: English
TARGET READER: Theoretical Computer Scientists, Information
Theorists

Publishes original papers in all areas of theoretical computer science
and computational aspects of information theory. Papers contributing
new results in active theoretical areas are particularly welcome.

INSTITUTE OF MATHEMATICAL STATISTICS BULLETIN
ISSN: 0146-3942

EDITOR: George P. H. Styan
                           Department of Math and Statistics
                           McGill University
                           Burnside Hall/1240 805 Ouest
                           Rue Sherbrooke St., W
                           Montreal, Quebec, CANADA, H3A 2K6
                           Tel: 1-(514)398-3845
                           FAX: 1-(514)398-3899
PUBLISHER: Institute of Mathematical Statistics
DATE FOUNDED: 1972   FREQUENCY: Bi-Monthly
PRICE: $40   MICROFORM: UMI
LANGUAGE: English CIRC: 4000
INDEXED/ABSTRACTED: Math R
TARGET READER: Statisticians, Probabilists

Publishes the complete programs of all IMS meetings, abstracts of all
invited and contributed papers, an international calendar of statistical
events as well as articles and news of interest to IMS members and to
statisticians and probabilists in general.

INTEGRAL EQUATIONS AND OPERATOR THEORY
ISSN: 0378-620X

EDITOR: I. Gohberg, Editorial Office
                           School of Mathematics
                           Tel Aviv University
                           Ramat Aviv, ISRAEL
MS REQUIREMENT: Original and one copy, 1½ spaced, 10-12
lines of English abstract .
PUBLISHER: Birkhaüser Verlag

DATE FOUNDED: 1978   FREQUENCY   Bi-Monthly
PRICE: $248   LANGUAGE: English
INDEXED/ABSTRACTED: Compumath, Math R
TARGET READER: Mathematicians, Physicists, Engineers,
Analysts

Publishes papers of current research in integral equations, operator
theory and related topics with emphasis on the linear aspects of the
theory. Reports on the full scope of current developments from abstract
theory to numerical methods and application to analysis, physics,
mechanics, engineering and others.

INTERNATIONAL JOURNAL OF ALGEBRA AND COMPUTATION
ISSN: 0218-1967

EDITOR: J. Rhodes
                             Department of Math
                             University of California
                             Berkeley, CA 94720
MS REQUIREMENT: Duplicate, dbl spaced, abstract not to be
more than 300 words, 1980 Mathematics Subject Classification
Numbers (1985 Rev.)
PUBLISHER: World Scientific Publishing Co.
DATE FOUNDED: 1991   FREQUENCY: Quarterly
PRICE: $185   LANGUAGE: English
TARGET READER: Mathematicians interested in discrete
mathematics

Publishes papers on algebra, combinatorics, graph theory as well as the
computational aspects of those subjects.

INTERNATIONAL JOURNAL OF APPROXIMATE REASONING
ISSN: 0888-613X

EDITOR: James C. Bezdek

Computer Science Department
University of West Florida
Pensacola, FL 32514
MS REQUIREMENT: Triplicate, dbl spaced, 200-300 word
abstract, 5-10 key words.
PUBLISHER: Elsevier
DATE FOUNDED: 1987 FREQUENCY: Bi-Monthly
PRICE: $178 Back issues available
INDEXED/ABSTRACTED: Art Intell Abstr, Math R
TARGET READER: Computer Scientists, Mathematicians,
Statisticians, Engineers, Economists

Dedicated to the dissemination of research results from the field of
approximate reasoning and its applications, with emphasis on the design
and implementation of intelligent systems of real-world applications.
Publishes refereed research reports, surveys, historical and critical
reviews, short notes and communications, problems, news about the
North American Fuzzy Information Processing Society, etc.

INTERNATIONAL JOURNAL OF COMPUTATIONAL GEOMETRY AND APPLICATIONS
ISSN: 0218-1959

PUBLISHER: World Scientific Publishing Co.
DATE FOUNDED: 1991 FREQUENCY: Quarterly
PRICE: $175 LANGUAGE: English
TARGET READER: Computer Scientists

Publishes papers on computational geometry and computer graphics.

INTERNATIONAL JOURNAL OF COMPUTER MATHEMATICS
ISSN: 0020-7160

EDITOR: David J. Evans

Department of Computer Studies
Loughborough University of Technology
Loughborough
Leics, LE11 3TU, ENGLAND
MS REQUIREMENT: Dbl spaced, 100-150 word summary,
non-English papers need up to a 300 word English abstract, key
words and computing reviews categories
PUBLISHER: Gordon and Breach Science Publishers
DATE FOUNDED: 1964   FREQUENCY: 16/yr.
PRICE: Various   MICROFORM: MIM
                       send request   LANGUAGE: English
                       to publisher   (preferred), French and German
                                                 (with English abstracts)
INDEXED/ABSTRACTED: Appl Mech Rev, Compumath, Comput
Rev, Eng Ind, Math R, Sci Abstr, Sci Cit Ind
TARGET READER: Applied Mathematicians, Computer Scientists

Section A publishes papers concerning research and development in
computer systems and the theory of programming languages. Section
B concerns mathematical techniques that are of interest to computer
users in the fields of numerical analysis, mathematical software, discrete
mathematics, computational geometry and graphics, pattern recognition,
operations research, and applied mathematics in general.

*INTERNATIONAL JOURNAL OF GAME THEORY**
ISSN: 0020-7276

PUBLISHER: Springer-Verlag
DATE FOUNDED: 1971   FREQUENCY: Quarterly
PRICE: $246.16 LANGUAGE: English
INDEXED/ABSTRACTED: Compt Control Abstr, Elect Electron
Abstr, Math R, Phys Abstr
TARGET READER: Mathematicians interested in game theory,
combinatorics, graph theory

Publishes original articles on the theory of games and its applications.

INTERNATIONAL JOURNAL OF MATHEMATICAL EDUCATION IN SCIENCE AND TECHNOLOGY
ISSN: 0020-739X

EDITOR: D. Walker
                           Department of Mathematical Sciences
                           University of Technology
                           Loughborough, LE11 3TU, ENGLAND
MS REQUIREMENT: Triplicate, dbl spaced, maximum 200 word summary.
PUBLISHER: Taylor and Francis
DATE FOUNDED: 1970   FREQUENCY: Bi-Monthly
PRICE: (Inst) $280   MICROFORM: UMI
               (Indv) $140   LANGUAGE: English
INDEXED/ABSTRACTED: CIJE, Educ Ind, Educ Tech Abstr, Math R, Sci Abstr
TARGET READER: Mathematicians, Mathematics Educators in schools, colleges, polytechnics, and universities

Provides a medium from which a wide range of experiences in mathematical education can be presented, assimilated and eventually adapted to everyday needs in schools, colleges, polytechnics, universities, industry and commerce. The important features are: mathematical models arising from real situations, the use of computers, non-teaching aids and techniques and discussions on methods of widening applications throughout science and technology.

INTERNATIONAL JOURNAL OF MATHEMATICS
ISSN: 0129-167X

EDITOR: Andrew Casson
                           Department of Math
                           University of California

Berkeley, CA 94720
MS REQUIREMENT: Duplicate, dbl spaced, maximum 300 word
abstract, references, tables
PUBLISHER: World Scientific Publishing Co.
DATE FOUNDED: 1990   FREQUENCY: Bi-Monthly
PRICE: $250   LANGUAGE: English
TARGET READER: Mathematicians in all fields

Publishes mathematical papers in all fields.

INTERNATIONAL JOURNAL OF MATHEMATICS AND MATHEMATICAL SCIENCES
ISSN: 0161-1712

EDITOR: Lokenath Debnath
                       International Journal of Mathematics and
                       Mathematical Sciences
                       University of Central Florida
                       Orlando, FL 32816
MS REQUIREMENT: Duplicate, dbl spaced, abstract no more
than 200 words, keywords and phrases, 1980 Mathematics Subject
Classification Codes
PUBLISHER: International Journal of Mathematics and
                       Mathematical Sciences
DATE FOUNDED: 1978   FREQUENCY: Quarterly
PRICE: $80   CIRC: 200
LANGUAGE: English Back Issues Available
INDEXED/ABSTRACTED: Appl Mech Rev, Math R, Zent Math
TARGET READER: Mathematicians, Physicists, Biologists

Publishes original research papers, research notes, research-expository
and survey articles with emphasis on unsolved problems and open
questions in Mathematics and Mathematical Sciences. Subjects such as
pure and applied mathematics, mathematical physics, theoretical
mechanics, probability and mathematical statistics, theoretical biology
are all accepted.

INTERNATIONAL JOURNAL OF NON-LINEAR MECHANICS
ISSN: 0020-7462

EDITOR:	William A. Nash
	College of Engineering
	University of Mass.
	Amherst, MA 01003
PUBLISHER:	Pergamon Press

DATE FOUNDED: 1966 FREQUENCY: Bi-Monthly
PRICE: $485 CIRC: 1200
MICROFILM: MIM, LANGUAGE: English, French,
        UMI, Pergamon   German
INDEXED/ABSTRACTED: Appl Mech Rev, Cam Sci Abstr, Curr
Cont, Eng Ind, Math R, Sci Abstr
TARGET READER: Mathematicians, Engineers, Physicists

Provides a specific medium for dissemination of research results in the various areas of theoretical and applied mechanics of solids and fluids as well as control theory.

INTERNATIONAL STATISTICAL REVIEW
ISSN: 0306-7734

HISTORY: *Revue de l'Institut International de Statistique*
        [ISSN: 0020-8779] V1-V39 (1933-1971)

EDITOR:	David R. Brillinger
	Department of Statistics
	University of California
	Berkeley, CA 94720
SEND MS TO:	International Statistical Institute
	428 Princes Beatrixlaan
	2270 AZ Voorburg,
	THE NETHERLANDS

MS REQUIREMENT: Four copies, dbl spaced, English and French summary respectively, key words in English, follow English summary.

PUBLISHER: International Statistical Institute
DATE FOUNDED: 1972   FREQUENCY: 3/yr.
PRICE: $50   CIRC: 3000
LANGUAGE: English, French
INDEXED/ABSTRACTED: Compumath, Curr Cont, Math R, Sci Abstr, Sci Cit Ind, Stat Theory Meth Abstr
TARGET READER: Statisticians, Probabilists

Provides a comprehensive view of work in statistics, over the whole spectrum of the statistical profession and including the most relevant aspects of probability. Publishes original research papers, survey papers and papers on the history of statistics and probability. Short reports on recent activities in the statistical world are accepted.

INVENTIONES MATHEMATICAE
ISSN: 0020-9910

PUBLISHER: Springer-Verlag
DATE FOUNDED: 1966   FREQUENCY: Monthly
PRICE: $1914.21 REPRINT: ISI
MICROFORM: UMI   LANGUAGE: English, French, German
INDEXED/ABSTRACTED: Compumath, Curr Cont, Math R, Sci Cit Ind, Zent Math
TARGET READER: Mathematicians in all fields

Publishes leading fundamental and authoritative papers on mathematics.

INVERSE PROBLEMS
ISSN: 0266-5611

SEND MS TO: Managerial Editor
Inverse Problems
IOP Publishing, Ltd.
Techno House, Redcliffe Way

Bristol BS1 6NX, ENGLAND
MS REQUIREMENT: Triplicate, International Classification for
Physics Numbers, abstract, accepts articles in TEX.
PUBLISHER: IOP Publishing, Ltd.
DATE FOUNDED: 1985   FREQUENCY: Bi-Monthly
PRICE: $550   CIRC: 550
LANGUAGE: English, French, German
INDEXED/ABSTRACTED: Compumath, Curr Cont, Deep Sea
Res & Oceanogr Abstr
TARGET READER: Pure and Applied Mathematicians, Physicists,
Researchers in geophysics, optics, radar, acoustics, communication
theory, and signal processing

Aims to combine theoretical and mathematical papers on inverse
problems with numerical and practical approaches to their solution.
Publishes contributions in two main categories: Letters to the Editor
and research papers. Papers are reports of original research work and
are not normally more than 8500 words in length. Letters are brief
communications that are both timely and important and are not normally
more than 2500 words in length.

ISRAEL JOURNAL OF MATHEMATICS
ISSN: 0021-2172

HISTORY: *Bulletin of the Research Council of Israel,*
                      Section A: Mathematics, Physics, and Chemistry
                      [ISSN: 0578-901X] V1-V6 (1955-1957)
                      Bulletin of the Research Council of Israel,
                      Section F: Mathematics and Physics
                      [ISSN: 0366-2799] V7-V10 (1957-1962)

SEND MS TO: The Editor
                              Israel Journal of Mathematics
                              Department of Math
                              The Hebrew University of Jerusalem
                              Jerusalem, ISRAEL
MS REQUIREMENT: Duplicate, triple spaced.

PUBLISHER: Weizmann Science Press of Israel
DATE FOUNDED: 1963 FREQUENCY: Monthly
PRICE: $240 CIRC: 950
LANGUAGE: English
INDEXED/ABSTRACTED: Compumath, Curr Cont, Ind Sci Rev, Math R, Sci Abstr, Sci Cit Ind
TARGET READER: Mathematicians in all fields.

Publishes short and medium length research papers in mathematics.

JAPAN JOURNAL OF INDUSTRIAL AND APPLIED MATHEMATICS
ISSN: 0916-7005

HISTORY: *Japan Journal of Applied Mathematics*
[ISSN: 0910-2043] V1-V7 (1984-1990)

MANAGING Masaya Yamaguti
EDITOR: Department of Math and Informatics
Ryukoku University
Seta, Ohtsu 520-21, JAPAN
MS REQUIREMENT: Triplicate, dbl spaced, 150 word abstract,
5 or less key words
PUBLISHER: Kinokuniya Co., Ltd.
DATE FOUNDED: 1991 FREQUENCY: 3/yr.
PRICE: Direct Order LANGUAGE: English
TARGET READER: Applied Mathematicians, Social Scientists,
Natural Scientists, Engineering Scientists

Provides an international forum for expression of new ideas as well as for exposition of original research and excellent case studies in the applied mathematics related to natural, social, and engineering sciences.

JAPANESE JOURNAL OF MATHEMATICS (New Series)
ISSN: 0075-3432

HISTORY: *Japanese Journal of Mathematics*
                  [ISSN: 0289-2316] V1-V42 (1924-1974)

EDITOR: Shoro Araki
                  Osaka City University
                  Osaka, JAPAN
PUBLISHER: Kinokuniya Co.
DATE FOUNDED: 1975 FREQUENCY: Semi-Annually
PRICE: $192 LANGUAGE: English, French,
                                       German
INDEXED/ABSTRACTED: Math R,
TARGET READER: Mathematicians in all fields.

Publishes research papers in all fields of mathematics.

JOURNAL D'ANALYSE MATHEMATIQUE
ISSN: 0021-7670

SEND MS TO: The Editor
                      Journal d'Analyse Mathematique
                      Department of Math
                      Bar-Ilan University
                      52900 Ramat-Gan, ISRAEL
PUBLISHER: Weizmann Science Press of Israel
DATE FOUNDED: 1951 FREQUENCY: Semi-Annually
PRICE: $120 CIRC: 750
LANGUAGE: English, French
INDEXED/ABSTRACTED: Compumath, Curr Cont, Ind Sci Rev,
Math R, Sci Cit Ind
TARGET READER: Analysts

Publishes original papers in the field of mathematical analysis.

JOURNAL DE MATHEMATIQUES PURES ET APPLIQUEES
ISSN: 0021-7824

HISTORY: *Annales de Mathematiques Pures et Appliquees,*
                T1-T22 (1810-1832)

SEND MS TO: Jacques-Louis LIONS
                    Secretaire, College de France
                    Place Marcelin-Berthelot
                    75231 Paris Cedex 05
PUBLISHER: Gauthier-Villars
DATE FOUNDED: 1836   FREQENCY: Quarterly
PRICE: $268.04   MICROFORM: UMI
CIRC: 1000   REPRINT: UMI
LANGUAGE: English, French
INDEXED/ABSTRACTED: Appl Mech Rev, Compumath, Curr
Cont, Ind Sci Rev, Math R, Sci Cit Ind
TARGET READER: Pure and applied mathematicians

Publishes research papers in both pure and applied mathematics.

JOURNAL FOR RESEARCH IN MATHEMATICS EDUCATION
ISSN: 0021-8251

EDITOR: Frank K. Lester, Jr.
                    Journal for Research in Mathematics
                    Education
                    Indiana University
                    Bloomington, IN 47405
MS REQUIREMENT: Six copies, dbl spaced, 100 word abstract
on a separate page
PUBLISHER: National Council of Teachers of
                    Mathematics
DATE FOUNDED: 1970   FREQUENCY: 5/yr.
PRICE: $30   LANGUAGE: English
INDEXED/ABSTRACTED: CIJE, Educ Ind
TARGET READER: Mathematics Educators

Devoted to the interests of teachers of mathematics and mathematics
education at all levels. Publishes high-quality manuscripts, such as

reports of research, including experiments, case studies, articles about research, including literature reviews and theoretical analyses, brief reports of research as well as critiques of articles and books.

JOURNAL FÜR DIE REINE UND ANGEWANDTE MATHEMATIK
ISSN: 0075-4102

EDITOR: Simon K. Donaldson
                         Math Institute
                         24-29 St. Giles'
                         Oxford, OX1 3LB ENGLAND
MS REQUIREMENT: Duplicate
PUBLISHER: Walter de Gruyer
DATE FOUNDED: 1826   FREQUENCY: 10/yr.
PRICE: $1598.40 MICROFORM: UMI
REPRINT: UMI   LANGUAGE: English, French,
                                                     German
INDEXED/ABSTRACTED: Compumath, Curr Cont, Math R, Sci Cit Ind, Zent Math
TARGET READER: Mathematicians in all fields.

Publishes only original mathematical works.   Book reviews and mathematical survey articles are not included.

JOURNAL OF ALGEBRA
ISSN: 0021-8693

EDITOR: Walter Feit
                         Yale University
                         Box 2155/Yale Station
                         New Haven, CT 06520
MS REQUIREMENT: Duplicate, dbl spaced
PUBLISHER: Academic Press
DATE FOUNDED: 1964   FREQUENCY: Monthly
PRICE: $1195.50 LANGUAGE: English

Back Issues Available
INDEXED/ABSTRACTED: Compumath, Curr Cont, Ind Sci Rev,
Math R, Sci Cit Ind
TARGET READER: Algebraists

Presents articles concerning original research in the field of algebra,
and related research areas that have a significant bearing on
algebra.

JOURNAL OF ALGORITHMS
ISSN: 0196-6774

EDITOR: Zvi Galil
                           Department of Computer Science
                           Columbia University
                           New York City, NY 10027
MS REQUIREMENT: Duplicate, dbl spaced, 150 word abstract
PUBLISHER: Academic Press
DATE FOUNDED: 1980   FREQUENCY: Quarterly
PRICE: $145   LANGUAGE: English
INDEXED/ABSTRACTED: Compumath, Elect Electron Abstr,
Int Abstr Oper Res, Math R, Sci Abstr
TARGET READER: Mathematicians, Computer Scientists

Presents papers on algorithms that are inherently discrete and finite and
that have some definite mathematical content in a natural way, either in
their objective or in their analysis. Also includes new algorithms and
data structures, new analysis or comparisons of known algorithms,
complexity studies and sharply focused review articles of subject areas
that are currently active. The journal also includes a problem section.

JOURNAL OF APPLIED MATHEMATICS AND
MECHANICS
ISSN: 0021-8928

Translation of *Prikladnaia Matematika i Mekhanika*

TRANSLATION	G. Herrmann
EDITOR:	Department of Applied Mechanics
	Stanford University
	Stanford, CA 94305

PUBLISHER:	Pergamon Press		
DATE FOUNDED:	1958	FREQUENCY:	6/yr.
PRICE:	$835	CIRC:	1200
LANGUAGE:	English	MICROFORM:	MIM, UMI,
			Pergamon

INDEXED/ABSTRACTED: Appl Mech Rev, Compumath, Curr
Cont, Eng Ind, Math R
TARGET READER: Mathematicians, Physicists, Engineers

Publishes papers with high-level mathematical investigations of modern
physical and mechanical problems and reports current progress in this
field.

JOURNAL OF APPLIED MATHEMATICS AND STOCHASTIC ANALYSIS*
ISSN: 1048-9533

| HISTORY: | *Journal of Applied Mathematics and Simulation* |
| | [ISSN: 0893-5688] V1-V2 (1987-1989) |

PUBLISHER:	University of Pittsburgh at Bradford	
DATE FOUNDED:	1990	FREQUENCY:	Quarterly
PRICE:	$100	LANGUAGE:	English

INDEXED/ABSTRACTED: Math R
TARGET READER: Applied Mathematicians, Analysts, Statisticians

Publishes papers on mathematical models and computer simulation.

JOURNAL OF APPLIED PROBABILITY
ISSN: 0021-9002

EDITOR: C. C. Heyde
                           Department of Probability
                           and Statistics
                           The University
                           Sheffield, S3 7RH, ENGLAND
MS REQUIREMENT: Dbl spaced, 4-10 line abstract, keywords
PUBLISHERS: Applied Probability Trust
                           London Mathematical Society
DATE FOUNDED: 1964   FREQUENCY: Quarterly
PRICE: (Inst) $150   CIRC: 1500
            (Indv) $66   LANGUAGE: English, French
INDEXED/ABSTRACTED: Biol Abstr, Curr Cont, Math R, Ref
Zh, Sci Abstr, Sci Cit Ind, Stat Theory Meth Abstr
TARGET READER: Probabilists, Biologists, Physicists

Publishes papers in two catagories: 1) Research papers not exceeding
20 printed pages; 2) Short communications of a few pages in the nature
of notes or brief accounts of work in progress. Research papers and
notes are on applications of probability theory to the biological,
physical, social, and technological sciences.

JOURNAL OF APPLIED STATISTICS
ISSN: 0266-4763

HISTORY: *Bulletin in Applied Statistics,*
               V1-V10 (1974-1983)

EDITOR: Gopal K. Kanji
                           Department of Applied
                           Statistics & Operational Research
                           Sheffield City Polytechnic
                           Pond Street
                           Sheffield S1 1WB, ENGLAND
MS REQUIREMENT: Duplicate, about 1500-3000 words
PUBLISHER: Carfax Publishing Co.
DATE FOUNDED: 1984   FREQUENCY: 3/yr.
PRICE: $208   CIRC: 700
Back Volumes Available

INDEXED/ABSTRACTED: Curr Ind Stat, J Cont Quant Meth
TARGET READER: Statisticians, Natural and Social Scientists

Provides a forum for applied statisticians across a wide range of
disciplines, such as business, computing, economics, ecology, education,
management, and medicine.

JOURNAL OF APPROXIMATION THEORY
ISSN: 0021-9045

EDITOR:	Paul Nevai
	Ohio State University
	Columbus, OH 43210-0341
EDITORIAL	Seventh Floor
OFFICE:	1250 6th Avenue
	San Diego, CA 92101
MS REQUIREMENT:	Triplicate, prefer triple spaced, brief
abstract	
PUBLISHER:	Academic Press
DATE FOUNDED:	1968 FREQUENCY: Monthly
PRICE:	$452 LANGUAGE: English,German

INDEXED/ ABSTRACTED: Compumath, Curr Cont, Ind Sci Rev,
Math R, Sci Abstr, Sci Cit Ind
TARGET READER: Approximation Theorists

Devoted to pure and applied approximation theory and related areas.
Covers basic theoretical aspects and application of approximation
theory.

JOURNAL OF CLASSIFICATION
ISSN: 0176-4268

EDITOR:	Phipps Arabie
	Graduate School of Management
	Rutgers University
	92 New Street
	Newark, NJ 07102-1895

PUBLISHER: Springer-Verlag
DATE FOUNDED: 1984   FREQUENCY: Semi-Annually
PRICE: $78   LANGUAGE: English
INDEXED/ABSTRACTED: Biol Abstr, Compumath, Curr Cont,
Math R, Zent Math
TARGET READER: Mathematicians, Social Scientists, Life
Scientists, Computer Researchers

Publishes original papers in the field of classification, numerical
taxonomy, multidimensional scaling and other ordination techniques,
clustering, tree structures and other network models, as well as
associated models and algorithms for fitting them. Contributions will
represent disciplines such as statistics, psychology, biology, information
retrieval, anthropology, archeology, astronomy, business, chemistry,
computer science, economics, engineering, geography, geology,
linguistics, mathematics, medicine, political science, psychiatry,
sociology. The journal has four sections in each issue: articles, short
notes and comments, software abstracts, and book reviews.

JOURNAL OF COMBINATORIAL MATHEMATICS AND COMBINATORIAL COMPUTING
ISSN: 0835-3026

EDITOR: W. D. Wallis
                        Department of Math
                        Southern Illinois University
                        Carbondale, IL 62901-4408
PUBLISHER: Charles Babbage Research Centre
DATE FOUNDED: 1987   FREQUENCY: Semi-Annually
PRICE: $42   LANGUAGE: English
Back Issues Available
INDEXED/ABSTRACTED: Math R
TARGET READER: Graph Theorists, Analysts, Computer
Scientists, Combinatorialists

Publishes papers on combinatorics and graph theory.

JOURNAL OF COMBINATORIAL THEORY - Series A
ISSN: 0097-3165

HISTORY: *Journal of Combinatorial Theory*
                  [ISSN 0021-9800] V1-V9 (1966-1970)

EDITOR: Basil Gordon
                  Department of Math
                  UCLA, Los Angeles, CA 90024
MS REQUIREMENT: Duplicate, prefer triple spaced, brief
abstract
PUBLISHER: Academic Press
DATE FOUNDED: 1971   FREQUENCY: Bi-Monthly
PRICE: $360   LANGUAGE: English
INDEXED/ABSTRACTED: Compumath, Curr Cont, Ind Sci Rev,
Math R, Sci Cit Ind
TARGET READER: Combinatorialists, Coding Theorists

Publishes original mathematical articles dealing with theoretical and
physical aspects of the study of finite and discrete structures in all
branches of science. Series A: Devoted mainly to structures, designs,
and applications of combinatorics.

JOURNAL OF COMBINATORIAL THEORY - Series B
ISSN: 0095-8956

HISTORY: *Journal of Combinatorial Theory*
                  [ISSN 0021-9800] V1-V9 (1966-1970)

EDITOR: Adrian Bondy
                  Department of Combinatorics
                  and Optimization, University of Waterloo
                  Waterloo, Ontario, CANADA N2L 3G1
MS REQUIREMENT: Duplicate, prefer triple spaced, brief
abstract
PUBLISHER: Academic Press
DATE FOUNDED: 1971   FREQUENCY: Bi-Monthly
PRICE: $318   LANGUAGE: English

INDEXED/ABSTRACTED: Compumath, Curr Cont, Ind Sci Rev, Math R, Sci Cit Ind
TARGET READER: Combinatorialists, Coding Theorists

Publishes original mathematical articles dealing with theoretical and physical aspects of the study of finite and discrete structures in all branches of science. Series B: Devoted mainly to graph theory and matroid theory.

JOURNAL OF COMBINATORICS, INFORMATION OF SYSTEM SCIENCES
ISSN: 0250-9628

EDITOR: Bhu Dev Sharma
                         Department of Math
                         P. O. Box 56B, Xavier University of LA
                         New Orleans, LA   70125
MS REQUIREMENT: Triplicate, no more than a 100 word abstract
PUBLISHER: Forum for Interdisciplinary Mathematics
DATE FOUNDED: 1976   FREQUENCY: Quarterly
PRICE: $60   CIRC: 200
LANGUAGE: English,   PUBL CHRGS: $12/Page
                  French, German
INDEXED/ABSTRACTED: Cyb Abstr, Math R
TARGET READER: Mathematicians, System Analysts, Information Theorists, Combinatorialists

Publishes original papers and survey articles on important advances, carrying theoretical results of mathematical orientation in one of the major areas: 1) Combinatorics; 2) Information Theory; 3) Mathematical System Theory. Book reviews, dissertation abstracts and comments on some already published papers are accepted.

JOURNAL OF COMPLEXITY
ISSN: 0885-064X

EDITOR: Joseph F. Traub
                           Computer Science Department
                           Columbia University
                           New York, NY 10027
MS REQUIREMENT: Quadruplicate, dbl spaced, short abstract
PUBLISHER: Academic Press
DATE FOUNDED: 1985   FREQUENCY: Quarterly
PRICE: $111   LANGUAGE: English
INDEXED/ABSTRACTED: Comput Control Abstr, Elect Electron
Abstr, Math R, Phys Abstr.
TARGET READER: Applied Mathematicians, Computer Scientists

Publishes original research papers that contain substantial mathematical
results on complexity as broadly conceived. Focuses on problems that
are approximately solved and for which optimal algorithms or
lower-bound results are available and provides major new algorithms or
important progress on upper bounds.

JOURNAL OF COMPUTATIONAL AND APPLIED MATHEMATICS
ISSN: 0377-0427

EDITOR: J. Wimp
                           Department of Math and
                           Computer Science, Drexel University
                           Philadelphia, PA 19104
MS REQUIREMENT: Triplicate, dbl spaced, list of key words,
algorithms must be written in Fortran, Algol 60, 68.
PUBLISHER: North-Holland
DATE FOUNDED: 1975   FREQUENCY: 15/yr.
PRICE: $828.65   LANGUAGE: English
INDEXED/ABSTRACTED: Appl Mech Rev, Cam Sci Abstr, ISI
Curr Cont, Math R
TARGET READER: Applied Mathematicians, Numerical Analysts,
Computer Scientists

Publishes original papers of high scientific standard in all areas of applied mathematics. Concentrates on the interaction between different fields of applied mathematics. The main interest is on papers describing new computational techniques for solving scientific problems.

JOURNAL OF COMPUTATIONAL MATHEMATICS
ISSN: 0254-9409

EDITOR:	Feng Kang (1989)
	Beijing. CHINA
PUBLISHER:	Science Press
	Copublished outside of China by VSP
DATE FOUNDED:	1983 FREQUENCY: Quarterly
PRICE:	$279.02 LANGUAGE: English

INDEXED/ABSTRACTED: Compumath, Curr Cont
TARGET READER: Researchers in computer science and mathematics

It is an international journal covering numerical methods, analysis and application. Publishes original research papers in all branches of modern computational mathematics, such as linear and nonlinear algebra, numerical optimization, numerical approximations, computational geometry, computational statistics, and probability, etc.

JOURNAL OF COMPUTER AND SYSTEM SCIENCES
ISSN: 0022-0000

EDITOR:	E. K. Blum, Math Department
	University of So. California
	Los Angeles, CA 90089
MS REQUIREMENT:	Duplicate, prefer triple spaced, short abstract
PUBLISHER:	Academic Press
DATE FOUNDED:	1967 FREQUENCY: Bi-Monthly
PRICE:	$348 LANGUAGE: English

INDEXED/ABSTRACTED: Biol Abstr, Compumath, Curr Cont,
Eng Ind, Math R, Sci Abstr, Sci Cit Ind
TARGET READER: Mathematicians, Computer Scientists

Publishes original research papers in computer science and system
science with emphasis on the relevant mathematical theory and its
application.

*JOURNAL OF CRYPTOLOGY**
ISSN: 0933-2790

PUBLISHER: Springer-Verlag
DATE FOUNDED: 1988   FREQUENCY: Quarterly
PRICE: $95   LANGUAGE: English
INDEXED/ABSTRACTED: Comput Control Abstr, Elect Electron
Abstr, Math R, Phys Abstr
TARGET READER: Mathematicians, Computer Scientists,
Cryptologists, Engineers

Provides original results in all areas of modern information security.
Topics include but are not limited to public key and conventional
algorithms   and   their   implementations,   cryptanalytic   attacks,
computational number theory and cryptographic protocols.

JOURNAL OF DIFFERENTIAL EQUATIONS
ISSN: 0022-0396

HISTORY: *Contributions to Differential Equations*
                   [ISSN: 0589-5839] V1-V3 (1963-1964)

EDITOR: Jack K. Hale
                   School of Math
                   Georgia Tech
                   Atlanta, GA 30332
MS REQUIREMENTS: Duplicate, prefer triple spaced
PUBLISHER: Academic Press
DATE FOUNDED: 1965   FREQUENCY: Monthly

PRICE: $708   LANGUAGE: English
INDEXED/ABSTRACTED: Comput Control Abstr, Elect Electron
Abstr, Math R, Phys Abstr
TARGET READER: Mathematicians, Engineers, Physicists, and
other Scientists who use differential equations as research tools

Publishes articles with the theory and the application of differential
equation. The articles are addressed not only to mathematicians, but
also to those engineers, physicists, and other scientists for whom
differential equations are valuable research tools.

JOURNAL OF DIFFERENTIAL GEOMETRY
ISSN: 0022-040X

SEND MS TO: Ms. Cindy Martin
                         Department of Math, Harvard University
                         Cambridge, MA 02138
MS REQUIREMENT: Typewritten, dbl spaced
PUBLISHER: Lehigh University
DIST BY: American Mathematical Society
DATE FOUNDED: 1967   FREQUENCY: Bi-Monthly
PRICE: (Inst) $250   CIRC: 950
                (Indv) $63   LANGUAGE: English, French,
Back Issues Available   German, Italian
INDEXED/ABSTRACTED: Compumath, Math R
TARGET READER: Mathematicians interested in differential
geometry, algebraic geometry

Publishes research papers in differential geometry and related subjects
such as differential equations, mathematical physics, algebraic geometry
and geometric topology.

JOURNAL OF DYNAMICS AND DIFFERENTIAL EQUATIONS
ISSN: 1040-7294

EDITOR: George R. Sell, School of Math

University of Minnesota
206 Church St., SE
Minneapolis, MN 55455
MS REQUIREMENT: Triplicate, dbl spaced, abstract less than
150 words, 4-5 key words
PUBLISHER: Plenum
DATE FOUNDED: 1989   FREQUENCY: Quarterly
PRICE: (Inst) $115   LANGUAGE: English
                (Indv) $50
INDEXED/ABSTRACTED: Curr Cont, Curr Math Publ, Math R
TARGET READER: Mathematicians interested in differential
equations, Biologist, Engineers, Physicists

Publishes original papers on the theory of the dynamics of differential
equations and their discrete analogs. Also publishes papers in other
areas of mathematics that have direct bearing on the dynamics of
differential equations.

JOURNAL OF ENGINEERING MATHEMATICS
ISSN: 0022-0833

EDITOR: H. K. Kuiken
                       Philips Research Laboratories
                       P. O. Box 80 000
                       5600 JA Eindhoven,
                       THE NETHERLANDS
SEND MS TO: Kluwer Academic Publishers
PUBLISHER: Kluwer Academic Publishers
DATE FOUNDED: 1967   FREQUENCY: Quarterly
PRICE: $194.50   LANGUAGE: English
MICROFORM: UMI
INDEXED/ABSTRACTED: Comput Control Abstr, Elect Electron
Abstr, Eng Ind, Math R, Phys Abstr
TARGET READER: Mathematicians, Engineers

Publishes original papers in various areas of the application of
mathematics to engineering science. Promotes the application of
mathematics to engineering problems and to stress the intrinsic unity of

the fundamental problems of different branches of engineering.

*JOURNAL OF FRACTIONAL CALCULUS**

EDITOR: Katsuyuki Nishimoto
                         2-13-10 Kaguike
                         Koriyama, Fukuskima-ken
                         JAPAN 963
MS REQUIREMENT: Original and two copies, dbl spaced,
abstract no more than 200 words.
PUBLISHER: Descartes Press Co.
DATE FOUNDED: 5/1992   FREQUENCY: Semi-Annually
PRICE: $95   LANGUAGE: English
PUBLICATION CHARGES: $28/page
TARGET READER: Mathematicians interested in differential
equations.

Publishes original contributions to pure and applied mathematics in the
subject including serendipities and related articles such as integral
transformations.

JOURNAL OF FUNCTIONAL ANALYSIS
ISSN: 0022-1236

EDITOR: Ralph S. Phillips
                         Department of Math
                         Stanford University
                         Stanford, CA 94305
MS REQUIREMENT: Duplicate, prefer triple spaced, abstract, list
of symbols
PUBLISHER: Academic Press
DATE FOUNDED: 1967   FREQUENCY: 16/yr.
PRICE: $832   LANGUAGE: English, French
INDEXED/ABSTRACTED: Compumath, Curr Cont, Ind Sci Rev,
Math R, Sci Abstr, Sci Cit Ind
TARGET READER: Analysts, Researchers interested in function

analytic character

Publishes original research papers of high quality from all branches of science, provided the core and flavor are of a functional analytic character. Also features applications and examples from other areas of mathematics, new developments in the field, and problems and challenges to functional analysis.

JOURNAL OF GEOMETRY
ISSN: 0047-2468

EDITORIAL OFFICE:	H. J. Kroll
	Institut für Geometrie der
	Technischen Universität Munchen
	Postfach 202420
	Ð 8000 Munchen 2, FRG
PUBLISHER:	Birkhaüser-Verlag

DATE FOUNDED: 1971   FREQUENCY: 6/yr.
PRICE: $198   CIRC: 1000
LANGUAGE: English, German
INDEXED/ABSTRACTED: Math R
TARGET READER: Mathematicians interested in geometry, Geometers

Publishes papers of current developments in geometry, particularly of recent results in foundations of geometry, geometric algebra, finite geometries, combinatorial geometry, and special geometries.

JOURNAL OF GEOMETRY AND PHYSICS*
ISSN: 0393-0440

EDITOR: Marco Modugno
PUBLISHER: Pitagora Editrice
DATE FOUNDED: 1984   FREQUENCY: Quarterly
PRICE: $120   LANGUAGE: English, French
Back Issues Available
INDEXED/ABSTRACTED: Math R

TARGET READER: Geometricians, Physicists

Publishes papers to promote interaction between geometry and physics,
which include articles on mathematical physics, pure geometry and
physics.

JOURNAL OF GRAPH THEORY
ISSN: 0364-9024

EDITOR: Fan Chung
                     Bell Communcations Research
SEND MS TO: The Managing Editor
                     Editorial Office of JGT, Bellcore
                     445 South St., Room 2P-391
                     Morristown, NJ 07960-1910
PUBLISHER: John Wiley
DATE FOUNDED: 1976   FREQUENCY: Bi-Monthly
PRICE: $230   MICROFORM: RPI
LANGUAGE: English
INDEXED/ABSTRACTED: Comput Control Abstr, Compumath,
Curr Cont, Elect Electron Abstr, Math R, Sci Cit Ind
TARGET READER: Graph Theorists, Computer Scientists,
Operations Reseachers, Communications Scientists

Covers a variety of topics in graph structures as well as algorithms with
theoretical emphasis. Includes related areas in combinatorics and the
interaction of graph theory with other mathematical sciences.

JOURNAL OF INTEGRAL EQUATIONS AND APPLICATIONS
ISSN: 0897-3962

HISTORY: *Journal of Integral Equations*
                [ISSN: 0163-5549] V1-V10 (1979-1985)

EDITOR: P. M. Anselone

		Math Department
		Oregon State University
		Corvallis, OR 97331
MS REQUIREMENT:		Duplicate, dbl spaced
PUBLISHER:		Rocky Mountain Mathematics
		Consortium
DATE FOUNDED:	1988	FREQUENCY: Quarterly
PRICE: (Inst)	$150	LANGUAGE: English
(Indv)	$50	PUBLICATION CHG: $35 to
		the Institution or the contract
		supporting the research

INDEXED/ABSTRACTED: Eng Ind, Eng Ind Bioeng Abstr, Eng
Ind Energ Abstr, Math R
TARGET READER: Mathematicians in analysis, Physicists,
Engineers

Publishes research papers and expository articles in the theory,
numerical analysis and application of integral equations. The scope and
methodologies will embrace classical and complex analysis methods,
functional analysis techniques and topological/geometric methods for
development of the theory of integral equations. Research papers will
not exceed twenty printed pages.

JOURNAL OF LOGIC AND COMPUTATION
ISSN: 0955-792X

EDITOR:	D. M. Gabbay
	Department of Computing
	Imperial College of Science
	Technology and Medicine
	180 Queen's Gate
	London, SW7 2BZ, ENGLAND
PUBLISHER:	Oxford University Press
DATE FOUNDED:	1990 FREQUENCY: Quarterly
PRICE:	$135 LANGUAGE: English

INDEXED/ABSTRACTED: Comput Control Abstr, Elect Electron
Abstr, Phys Abstr
TARGET READER: Computer Scientists, Logicians

Publishes technical scientific papers in virtually all aspects of Information Technology, from software engineering and hardware to programming and artificial intelligence. The journal aims to promote the growth of logic and computing.

JOURNAL OF LOGIC PROGRAMMING
ISSN: 0743-1066

EDITOR: J. L. Lassez
                            IBM-T. J. Watson Research Center
                            Yorktown Heights, NY 10598
MS REQUIREMENT: Five copies, dbl spaced, 200 word abstract
PUBLISHER: North-Holland
DATE FOUNDED: 1984   FREQUENCY: 8/yr.
PRICE: $278   MICROFORM: UMI
LANGUAGE: English   CIRC: 1000
INDEXED/ABSTRACTED: AI Abstract, Compumath,
Comput Rev, Eng Ind, Math R, Sci Abstr
TARGET READER: Logicians, Researchers, and Educators
concerned with the theory of computing, artificial intelligence and
programming languages, Software Engineers and Developers.

Publishes original research papers, survey and review articles, tutorial expositions, and historical studies in the area of logic programming. All aspects will be covered including theory and foundations, implementation issues, applications involving novel ideas and relationships with other programming methodologies.

JOURNAL OF MATHEMATICAL ANALYSIS AND APPLICATIONS
ISSN: 0022-247X

SEND MS TO   Editorial Office:
                            Seventh Floor
                            1250 6th Avenue
                            San Diego, CA 92101
MS REQUIREMENT: Original and two copies, prefer trpl

spaced, paper should be less than 30 pages
PUBLISHER: Academic Press
DATE FOUNDED: 1960   FREQUENCY: 18/yr.
PRICE: $1332   LANGUAGE: English
Back Issues Available
INDEXED/ABSTRACTED: Appl Mech Rev, Compumath, Curr
Cont, Eng Ind, Math R, Sci Abstr, Sci Cit Ind
TARGET READER: Analysts, Engineers, Biologists, Chemists

Publishes mathematical papers that treat classical analysis and its
numerous applications. Papers devoted to the mathematical treatment
of questions arising in physics, chemistry, biology, and engineering,
particularly those that stress analytical aspects and novel problems and
their solutions.

*JOURNAL OF MATHEMATICAL AND PHYSICAL SCIENCES**
ISSN: 0047-2557

EDITOR: P. Achuthan
PUBLISHER: Indian Institute of Technology
                              Madras, INDIA
DATE FOUNDED: 1967   FREQUENCY: Bi-Monthly
PRICE: (Inst) $50   LANGUAGE: English
                 (Indv) $20   CIRC: 300
INDEXED/ABSTRACTED: Appl Mech Rev, Curr Cont, Elect
Electron Abstr, Sci Abstr, Zent Math
TARGET READER: Mathematicians, Physicists

Publishes mathematical papers which are of broad interest in physical
sciences.

JOURNAL OF MATHEMATICAL BEHAVIOR
ISSN: 0732-3123

HISTORY: *Journal of Children's Mathematical Behavior*
                      [ISSN: 0160-0133] V1-V2 (1972-1979)

EDITOR: Robert B. Davis
                           Graduate School of Education
                           Rutgers University
                           10 Seminary Place
                           New Brunswick, NJ 08903
MS REQUIREMENT: Four copies, dbl spaced, abstract between
100-150 words
PUBLISHER: Åblex Publishing Corp.
DATE FOUNDED: 1980   FREQUENCY: 3/Yr.
PRICE: (Inst) $80   CIRC: 500
                (Indv) $35   LANGUAGE: English
INDEXED/ABSTRACTED: Math R, Psychol Abstr
TARGET READER: Mathematics Educators

Publishes papers which will assist in interventions that improve the
teaching and learning of mathematics and to help develop a deeper
understanding of how people learn and use mathematics.

JOURNAL OF MATHEMATICAL BIOLOGY
ISSN: 0303-6812

EDITOR: H. T. Banks
                           Center for Applied Mathematical
                           Sciences, DRB-306
                           University of Southern California
                           Los Angeles, CA 90089-1113
MS REQUIREMENT: Triplicate, abstract, 5 key words
PUBLISHER: Springer Verlag
DATE FOUNDED: 1974   FREQUENCY: 6/yr.
PRICE: $409   LANGUAGE: English
MICROFILM: UMI   REPRINTS: ISI
INDEXED/ABSTRACTED: Compumath, Curr Cont, Excerpta Med,
Math R, Sci Abstr, Zent Math
TARGET READER: Mathematicians, Biologists

Publishes papers in which mathematics is used to gain a better
understanding of biological phenomena; mathematical papers are

inspired by biological research and papers which yield new experimental data bearing on mathematical models.

JOURNAL OF MATHEMATICAL CHEMISTRY
ISSN: 0259-9791

EDITOR: P. G. Mezey
Department of Chemistry & Mathematics
University of Saskatchewan
Saskatoon, Saskatchewan
S7N 0W0, CANADA
FAX: (1-306) 966-4777
MS REQUIREMENT: Triplicate, abstract not to exceed 150 words
PUBLISHER: J. C. Baltzer, AG
DATE FOUNDED: 1987 FREQUENCY: Monthly
PRICE: $876.63 LANGUAGE: English
MICROFORM: Publisher
INDEXED/ABSTRACTED: Chem Abstr, Math R
TARGET READER: Chemists, Mathematicians

Publishes papers dealing with the novel and nontrivial application of any branch of mathematics to problems of chemical interest. Introduces new mathematical methods or techniques for the solution of chemical problems or develops new mathematical approaches or insights pertinent to any area of chemistry.

JOURNAL OF MATHEMATICAL ECONOMICS
ISSN: 0304-4068

EDITOR: Truman F. Bewley,
Cowles Foundation
for Research in Economics
P. O. Box 2125/Yale Station
New Haven, CT 06520-2125
MS REQUIREMENT: Dbl spaced, up to 100 word abstract, illustration should be sent in triplicate.

PUBLISHER: North-Holland
DATE FOUNDED: 1974   FREQUENCY: 6/yr.
PRICE: (Inst) $342.66   MICROFORM: UMI, RPI
                (Indv) $ 75   LANGUAGE: English only
INDEXED/ABSTRACTED: Bus Ind, Compumath, ISI Curr Cont,
Math R, SSCI
TARGET READER: Economists, Mathematicians

Provides a forum for work in economic theory which expresses
economic ideas using formal mathematical reasoning. Also provides
a forum for work which develops new mathematics inspired by
economic problems.

JOURNAL OF MATHEMATICAL PHYSICS
ISSN: 0022-2488

EDITOR: Lawrence C. Biedenharn, Jr.
SEND MS TO: Nancy J. Sipes
                        208C Physics Bldg., Duke University
                        Science Drive
                        Durham, NC 27706
MS REQUIREMENT: Triplicate, dbl spaced, abstract, index
article according to AIP's Physics and Astronomy Classification
Scheme-1990
PUBLISHER: American Institute of Physics
DATE FOUNDED: 1960   FREQUENCY: Monthly
PRICE: (Inst) $890   MICROFORM: AIP
                (Memb) $75   REPRINT: AIP

LANGUAGE: English
INDEXED/ABSTRACTED: Chem Abstr, Compumath, Curr Cont,
Eng Ind, Math R, Sci Abstr, Sci Cit Ind
TARGET READER: Mathematicians, Theoretical Physicists

Publishes mathematical physics articles on branches of mathematics that
are currently or potentially useful for the development of theoretical
physics.

JOURNAL OF MATHEMATICAL PSYCHOLOGY
ISSN: 0022-2496

EDITOR: Thomas S. Wallsten
                                 University of North Carolina
                                 Chapel Hill, NC 27514
PUBLISHER: Academic Press
DATE FOUNDED: 1964   FREQUENCY: Quarterly
PRICE: $178   LANGUAGE: English
INDEXED/ABSTRACTED: Biol Abstr, Compumath, Curr Cont,
Math R, Psychol Abstr
TARGET READER: Mathematicians, Psychologists

Presents theoretical and empirical research in all areas of mathematical
psychology. Areas of special interest include fundamental measurement
as well as the development and experimental testing of psychological
process models.

JOURNAL OF MATHEMATICAL SOCIOLOGY
ISSN: 0022-250X

EDITOR: Patrick Doreian
                                 Department of Sociology
                                 University of Pittsburgh
                                 Pittsburgh, PA 15260
MS REQUIREMENT: Triplicate, dbl spaced, abstract no longer
than 120 words, encourage TEX form
PUBLISHER: Gordon and Breach
DATE FOUNDED: 1971   FREQUENCY: Irr
PRICE: Contact Publisher   LANGUAGE: English
MICROFORM: Publisher
INDEXED/ABSTRACTED: Compumath, Curr Cont, Math R,
Psychol Abstr, SSCI, Sci Abstr
TARGET READER: Applied Mathematicians, Sociologists

Publishes articles in all areas of mathematical sociology. It publishes
papers in areas of mutual interest to sociologists and other social and
behavioral scientists which may encourage fruitful connections between

sociology and other disciplines. Articles deal primarily with the use of mathematical models in social science, the logic of measurement, computers, and computer programming, applied mathematics, statistics or quantitative methodology as they make some contribution to the understanding of substantive social phenomena.

*JOURNAL OF MATHEMATICAL SYSTEMS, ESTIMATIONS, AND CONTROL**
ISSN: 1052-0600

EDITOR: Clyde F. Martin
                               Texas Technical University
                               Lubbock, TX 79409
MS REQUIREMENT: Triplicate, brief abstracts, key words,
AMS (1980) Subject Classification Numbers, accept AMS TEX or
LATEX
PUBLISHER: Birkhaüser-Verlag
DATE FOUNDED: 1991   FREQUENCY: Quarterly
PRICE: $164   LANGUAGE: $164
TARGET READER: Control Estimation Theorists, System
Analysts

Publishes mathematically sophisticated papers in the areas of systems, estimation and control. Novel applications of contemporary mathematics to problems in estimation and control, or the broad area of mathematical system theory are welcome.

JOURNAL OF MATHEMATICS OF KYOTO UNIVERSITY
ISSN: 0023-608X

HISTORY: *Memoirs of the College of Science.*
                Kyoto Imperial University,
                V1-V6 (1914-1923).
                Memoirs of the College of Science. Series A,
                V7-V31 (1923-1967).

issued by the University of Kyoto.
Memoirs of the College of Science.
University of Kyoto. *Series A-Mathematics,*
V26-V33 (1950-1961).

EDITORIAL OFFICE:	Kyoto University, Department of Math Oiwake-cho, Kitashirakawa Sakyo-ku, Kyoto-shi Kyoto-fu 606, JAPAN
PUBLISHER:	Kyoto University, Department of Math
DIST BY:	Kinokuniya Co., Ltd.
DATE FOUNDED:	1961 FREQUENCY: 3/yr.
PRICE:	$173 CIRC: 900
LANGUAGE:	English, French

INDEXED/ABSTRACTED: Compumath, Math R, Sci Cit Ind
TARGET READER: Mathematicians in all fields

Publishes both pure and applied mathematical research papers.

JOURNAL OF MATHEMATICS, TOKUSHIMA UNIVERSITY
ISSN: 0075-4293

HISTORY: *Journal of Science of the Gakugei Faculty,*
                    Tokushima University, V1 (1950)
                    Journal of the Gakugei College,
                    Tokushima University, V2 (1952)
                    Journal of Gakugei, Tokushima University
                    [ISSN: 0593-6957] V3-V15 (1953-1966)

PUBLISHER:	Tokushima University
DATE FOUNDED:	1967 FREQUENCY: Annual
PRICE:	Exchange Basis LANGUAGE: English

INDEXED/ABSTRACTED: Math R
TARGET READER: Mathematicians in all fields

Publishes papers on all aspects of mathematics.

JOURNAL OF MULTIVARIATE ANALYSIS
ISSN: 0047-259X

EDITOR: C. R. Rao
                 Center for Multivariate Analysis
                 Department of Statistics
                 123 Pond Laboratory
                 Penn State University
                 University Park, PA 16802
PUBLISHER: Academic Press
MS REQUIREMENT: Triplicate (including original), trpl
spaced, brief abstract, key words, AMS Classification Numbers
DATE FOUNDED: 1971   FREQUENCY: 8/yr.
PRICE: $400   LANGUAGE: English
Back Issues Available   preferred, French, German
INDEXED/ABSTRACTED:. Compumath, Curr Cont, Eng Ind, Ind
Sci Rev, Math R
TARGET READER: Analysts

Publishes research results in the general area of multivariate analysis.
Covers articles on fundamental theoretical aspects of the field as well
as on other aspects concerned with significant applications of new
theoretical methods.

JOURNAL OF NUMBER THEORY
ISSN: 0022-314X

EDITOR: Alan Woods
                      Math Department
                      Ohio State University
                      Columbus, OH 43210
MS REQUIREMENT: Duplicate, triple spaced, short abstract, list
of symbols
PUBLISHER: Academic Press
DATE FOUNDED: 1969   FREQUENCY: Monthly
PRICE: $378   LANGUAGE: English
INDEXED/ABSTRACTED: Compumath, Curr Cont, Ind Sci Rev,
Math R, Sci Abstr

TARGET READER: Algebraists, Number Theorists, Coding
Theorists

Publishes selected research papers representing the broad spectrum of
interest of contemporary number theory and allied fields.

JOURNAL OF OPERATOR THEORY
ISSN: 0379-4024

EDITOR: William B. Arveson
                             Department of Math
                             University of California
                             Berkeley, CA 94720
MS REQUIREMENT: Duplicate, dbl spaced
PUBLISHER: Institute of Mathematics of the
                             Romanian Academy
DIST BY: American Mathematical Society
DATE FOUNDED: 1979   FREQUENCY: 4/yr.
PRICE: (Inst) $88   LANGUAGE: English, French
               (Indv) $56   German, Russian
Back Issues Available from AMS
INDEXED/ABSTRACTED: Compumath, Curr Cont, Sci Cit Ind
TARGET READER: Mathematicians in the field of operator theory

Publishes significant research articles in all areas of operator theory.

JOURNAL OF OPTIMIZATION THEORY AND APPLICATIONS
ISSN: 0022-3239

EDITOR: A. Miele
                             Aero-Astronautics Group
                             230 Ryon Bldg., Rice University
                             Houston, TX 77251-1892
MS REQUIREMENT: Triplicate, trpl spaced, 50-100 word

abstract, 4-5 words
PUBLISHER: Plenum
DATE FOUNDED: 1967   FREQUENCY: Monthly
PRICE: $725   LANGUAGE: English
INDEXED/ABSTRACTED: Appl Mech Rev., Compumath, Curr
Cont, Math R, Sci Abstr, Sci Cit Ind
TARGET READER: Mathematicians, Scientists, Economists,
Engineers

Publishes papers covering mathematical optimization techniques and
their applications to science and engineering. Covers five types of
contributions: Survey Papers, Contributed Papers, Technical Notes,
Technical Comments, and Book Reviews.

*JOURNAL OF PARTIAL DIFFERENTIAL EQUATIONS**
ISSN: 1000-940X

PUBLISHER: Pergamon Press
DATE FOUNDED: 1988   FREQUENCY: Quarterly
PRICE: $160   LANGUAGE: English
INDEXED/ABSTRACTED: Math R
TARGET READER: Engineers, Physicists, Mathematicians
particularly interested in partial differential equations

Publishes original research papers in both the theory and application of
partial differential equations. Encourages national and internatioal
exchange between the area of partial differential equations and
engineering, physics and other mathematical sciences.

JOURNAL OF PURE AND APPLIED ALGEBRA
ISSN: 0022-4049

EDITOR: P. J. Freyd
                                 Department of Math
                                 University of Pennsylvania
                                 Philadelphia, PA 19104

MS REQUIREMENT: Duplicate, dbl spaced, short abstract
PUBLISHER: North-Holland
DATE FOUNDED: 1971   FREQUENCY: 21/yr.
PRICE: $1088.25   MICROFORM: RPI
LANGUAGE: English
INDEXED/ABSTRACTED: Curr Cont, ISI Curr Cont, Math R
TARGET READER: Algebraists, Algebraic Topologists, Category
Theorists

Concentrates on that part of algebra likely to be of a general
mathematical interest: algebraic results with immediate applications,
and the development of algebraic theories of sufficiently general
relevance to allow for future applications.

JOURNAL OF RECREATIONAL MATHEMATICS
ISSN: 0022-412X

EDITOR: Joseph S. Madachy
                       4761 Rigger Road
                       Kettering, OH 45440
PUBLISHER: Baywood Publishing Co.
DATE FOUNDED: 1968   FREQUENCY: 4/yr.
PRICE: (Inst) $70   LANGUAGE: English
              (Indv) $18.95
INDEXED/ABSTRACTED: Gen Sci Ind, Math R
TARGET READER: Undergraduate and Graduate mathematics
students, math enthusiasts

Contains thought-provoking, stimulating, wit-sharpening games, puzzles,
and articles.

JOURNAL OF SOVIET MATHEMATICS
ISSN: 0090-4104

HISTORY: Merge of *Problems in Mathematical Analysis,*
                       (1968-1972)
                       Seminars in Mathematics

[ISSN: 0080-8873], (1968-1972)
Progress in Mathematics
[ISSN: 0079-6433], (1968-1972)

PUBLISHER: Consultants Bureau
DATE FOUNDED: 1973   FREQUENCY: 30/yr.
PRICE: $2450   LANGUAGE: English
INDEXED/ABSTRACTED: Comput Inf Syst Abstr J, Math R, Zent
Math
TARGET READER: Mathematicians in all fields

Covers original papers in all branches of mathematics.

JOURNAL OF STATISTICAL COMPUTATION AND SIMULATION
ISSN: 0094-9655

EDITOR: Richard G. Krutchkoff
                                    Department of Statistics
                                    Virginia Polytechnic Institute
                                    and State University,
                                    Blacksburg, VA 24061
MS REQUIREMENT: Triplicate, dbl spaced, no more than 400
word abstract, key words
PUBLISHER: Gordon and Breach
DATE FOUNDED: 1972   FREQUENCY: Irr
PRICE: Varies   LANGUAGE: English
INDEXED/ABSTRACTED: Compumath, Curr Cont, Ind Sci Rev,
Math R, Sci Abstr
TARGET READER: Statisticians, Computer Scientists

Publishes significant and original works in areas of statistics which are
related to or dependent upon the computer. Fields covered include
computer algorithms related to probability or statistics, studies in
statistical inference by means of simulation techniques and
implementation of interactive statistical systems.

JOURNAL OF STATISTICAL PLANNING AND INFERENCE
ISSN: 0378-3758

EDITOR: S. S. Gupta
                          Department of Statistics,
                          Purdue University
                          West Lafayette, IN 47907
MS REQUIREMENT: Four copies, dbl spaced, up to 200 word
abstract, list of math symbols, key words and phrases, AMS
Classification Numbers
PUBLISHER: North-Holland
DATE FOUNDED: 1977   FREQUENCY: 9/yr.
PRICE: $525.84   MICROFORM: RPI
LANGUAGE: English
INDEXED/ABSTRACTED: Curr Ind Stat, ISI Curr Cont, Math R,
Stat Theory Meth Abstr
TARGET READER: Mathematicians, Statisticians, Probabilists,
Econometricians, Mathematical and Economic Libraries

Provides a common medium for the dissemination of significant
information in all branches of statistics, with particular emphasis on
statistical planning and the related areas of combinatorial mathematics
and probability theory.

JOURNAL OF SYMBOLIC COMPUTATION
ISSN: 0747-7171

EDITOR: Bruno Buchberger
                          Johannes Kepler Universität
                          Institüt für Mathematik
                          A-4040 Linz, AUSTRIA
PUBLISHER: Academic Press
DATE FOUNDED: 1985   FREQUENCY: Monthly
PRICE: (Inst) $308   LANGUAGE: English
              (Indv) $154
INDEXED/ABSTRACTED: Compumath, Curr Cont
TARGET READER: Mathematicians, Physicists, Theoretical

Computer Scientists, Numerical Analysts

Provides a forum for research in the algorithmic treatment of all types of symbolic objects: objects in formal languages and geometrical objects.

JOURNAL OF SYMBOLIC LOGIC
ISSN: 0022-4812

EDITOR:	Sy D. Friedman
	Department of Math, MIT
	Cambridge, MA 02139
MS REQUIREMENT:	Duplicate, dbl spaced
PUBLISHER:	Association for Symbolic Logic
DATE FOUNDED:	1936 FREQUENCY: Quarterly
PRICE:	$165 MICROFORM: UMI
CIRC:	2500 LANGUAGE: English, French, German

INDEXED/ABSTRACTED: Compumath, Curr Cont, Hum Ind, Math R, Phil Ind, Sci Abstr
TARGET READER: Logicians, Philosophers

Publishes research papers in symbolic logic and serves as a forum for the exchange of ideas among mathematicians, philosophers, and others interested in this field.

JOURNAL OF THE AMERICAN MATHEMATICAL SOCIETY
ISSN: 0894-0347

EDITOR:	Wilfried Schmid
	Department of Math
	Harvard University
	Cambridge, MA 02138
PUBLISHER:	American Mathematical Society
DATE FOUNDED:	1988 FREQUENCY: Quarterly
PRICE:	$126 LANGUAGE: English
(Inst Memb)	$101 CIRC: 950

       (Indv Memb) $76   Back Volumes Available
INDEXED/ABSTRACTED: Math R
TARGET READER: Mathematicians in all fields

Publishes research articles of the highest quality in all areas of pure and applied mathematics.

JOURNAL OF THE AMERICAN STATISTICAL ASSOCIATION
ISSN: 0162-1459

HISTORY: *Publications of the American Statistical Association*, V1-V12 (1888-1912)
              Quarterly Publications of the American Statistical Association, V13-V17 (1912-1921).

JOURNAL   Linda A. Dziobek
MANAGER: American Statistical Association
                    1429 Duke Street
                    Alexandria, VA 22314
PUBLISHER: American Statistical Association
DATE FOUNDED: 1922   FREQUENCY: Quarterly
PRICE: $125   CIRC: 17,172
LANGUAGE: English   MICROFORM: UMI, MIM
REPRINT: UMI
INDEXED/ABSTRACTED: Bus Ind, Compumath, Comput Rev, Curr Cont, Math R
TARGET READER: Statisticians, Economists, Business Executives, Engineers, Sociologists

Publishes papers on the application of statistical methods to practical problems, in the development of more useful methods, and in the improvement of basic statistical data.

JOURNAL OF THE AUSTRALIAN MATHEMATICAL SOCIETY
SERIES A: PURE MATHEMATICS AND STATISTICS

ISSN: 0263-6115

HISTORY: *Journal of the Australian Mathematical Society*
                 [ISSN: 0004-9735], V1-V18 (1959-1974)
                 Journal of the Australian Mathematical Society.
                 Series A: Pure Mathematics [ISSN: 0334-3316]
                 V19-V28 (1975-1979)

EDITOR: T. E. Hall
                              Math Department
                              Monash University
                              Clayton, Victoria 3168
MS REQUIREMENT: Duplicate, dbl spaced, up to 150 word
abstract, 1991 AMS Math Subject Classification, encourage papers in
AMS-TEX form
PUBLISHER: Australian Mathematical Society
                              Department of Mathematics
                              University of Queensland
                              Qld 4072, AUSTRALIA
DATE FOUNDED: 1980   FREQUENCY: Bi-Monthly
PRICE: $160   LANGUAGE: English
Back Issues Available
INDEXED/ABSTRACTED: Appl Mech Rev, Compumath, Math R
TARGET READER: Pure Mathematicians, Statisticians

Publishes research papers in pure mathematics and mathematical
statistics.

JOURNAL OF THE AUSTRALIAN MATHEMATICAL SOCIETY, SERIES B - APPLIED MATHEMATICS
ISSN: 0334-2700

HISTORY: *Journal of the Australian Mathematical Society*
                 [ISSN: 0004-9735], V1-V18 (1959-1974)

EDITOR: E. O. Tuck
                              Applied Mathematics Department

University of Adelaide
G.P.O. Box 498
Adelaide, S.A. 5001 AUSTRALIA

MS REQUIREMENT: Triplicate, dbl spaced, up to 250 word
abstract

PUBLISHER: Australian Mathematical Society
                   Department of Math
                   University of Queensland

DATE FOUNDED: 1975   FREQUENCY: Quarterly
PRICE: $97   LANGUAGE: English

Back Issues Avaialale

INDEXED/ABSTRACTED: Appl Meth Rev, Compumath, Math R,
Sci Abstr

TARGET READER: Applied Mathematicians

Publishes papers in any field of applied mathematics and related
mathematical sciences, excluding statistics.

JOURNAL OF THE FACULTY OF SCIENCE, THE UNIVERSITY OF TOKYO
SECTION IA MATHEMATICS
ISSN: 0040-8980

HISTORY: *Journal of the College of Science,*
              Imperial University, Japan.
              V1-V9 (1887-1897)
              Journal of the College of Science,
              Imperial University of Toyko.
              V9-V45 (1898-1925)
              Journal of the College of Science,
              Imperial University of Toyko. Section 1:
              Mathematics, Astronomy, Physics, Chemistry.
              V1-V5 (1925-1944)
              Journal of the College of Science,
              University of Toyko,
              V5-V16 (1949-1970)

EDITOR: Yukio Matsumoto

                        Department of Math
                        Faculty of Science
                        University of Tokyo
                        Bunkyo-ku, Tokyo 113, JAPAN
PUBLISHER: University of Tokyo,
                        Department of Math
ORDER FROM: Tokyo International
                        Import and Export Department
DATE FOUNDED: 1970   FREQUENCY: Irr
PRICE: Varies   CIRC: 850
LANGUAGE: English, French, German
INDEXED/ABSTRACTED: Chem Abstr, Math R, Sci Abstr
TARGET READER: Mathematicians in all fields

Publishes original mathematical research papers in all areas.

*JOURNAL OF THE KOREAN MATHEMATICAL SOCIETY**
ISSN: 0304-9914

PUBLISHER: Korean Mathematical Society
DATE FOUNDED: 1964   FREQUENCY: Semi-Annually
PRICE: $40   LANGUAGE: English
INDEXED/ABSTRACTED: Math R
TARGET READER: Mathematicians in all areas

Publishes papers on all aspects of mathematics.

JOURNAL OF THE LONDON MATHEMATICAL SOCIETY
ISSN: 0024-6107

EDITOR: I. N. Baker/G. D. James
                    Math Department, Imperial College
                    London SW7 2BZ ENGLAND
MS REQUIREMENT: Duplicate, dbl spaced, AMS Classification
Numbers

PUBLISHER: London Mathematical Society
PRINTED BY: Cambridge Mathematical Press
DATE FOUNDED: 1926   FREQUENCY: Bi-Monthly
PRICE: $440   MICROFORM: UMI
LANGUAGE: English
Back Issues Available
INDEXED/ABSTRACTED: Compumath, Math R, Zent Math
TARGET READER: Mathematicians in all fields

The Society publishes longer papers in the Proceedings and the shorter
papers in the Journal to promote mathematical knowledge. Publishes
research papers from a broad spectrum within mathematics, but with the
main emphasis on pure mathematics.

JOURNAL OF THE MATHEMATICAL SOCIETY OF JAPAN
ISSN: 0025-5645

HISTORY: *Proceedings. Series 1. V1- Series 2.*
                V9 (1884-1918)
                Proceedings of the Physico - Mathematical
                Society. Series 3. V1-V26 (1919-1944)

SEND MS TO: The Editorial Committee of the Journal
                        The Math Society of Japan
                        25-9-203, Hongo 4-chome,
                        Bunkyo-ku, Tokyo 113, JAPAN
MS REQUIREMENT: Duplicate, Dbl spaced
PUBLISHER: The Mathematical Society of Japan
DATE FOUNDED: 1948   FREQUENCY: Quarterly
PRICE: $197   LANGUAGE: English, French,
                                             German
INDEXED/ABSTRACTED: Compumath, Curr Cont, Jap Per Ind,
Math R, Sci Cit Ind
TARGET READER: Mathematicians in all fields

Publishes original mathematical research in all fields.

JOURNAL OF THE ROYAL STATISTICAL SOCIETY.
SERIES A: STATISTICS IN SOCIETY
ISSN: 0035-9238

HISTORY: *Journal of the Statistical Society of London,*
                  V1-V35 (1838-1872)
                Journal of the Statistical Society,
                  V36-V49 (1873-1886)
                Journal of the Royal Statistical Society
                  [ISSN: 0925-8385] V50-V110 (1887-1947)
                Journal of the Royal Statistical Society.
                Series A: General
                  [ISSN: 0035-9238] V111-V150 (1948-1987)

SEND MS TO: Executive Secretary
                   Royal Statistical Society
                   25 Enford Street
                   London, W1H 2BH ENGLAND
                   Tel: 071-723-5882
                   (or) 071-723-6165
MS REQUIREMENT: Five copies, dbl spaced, summary of 100
words, 5-6 key words
PUBLISHER: Royal Statistical Society
                   Basil Blackwell
DATE FOUNDED: 1988   FREQUENCY: 3/yr.
PRICE: $64.50   CIRC: 5700
MICROFORM: UMI   LANGUAGE: English
INDEXED/ABSTRACTED: Appl Mech Rev, Compumath, Curr
Cont, Math R, SSCI
TARGET READER: Statisticians

Publishes papers in general statistical interest.   Papers containing
mathematical expositions and are relevant to the practice of statistics are
included.

JOURNAL OF THE ROYAL STATISTICAL SOCIETY.
SERIES B: METHODOLOGICAL
ISSN: 0035-9246

HISTORY: *Supplement to the Journal of the Royal Statistical Society*, V1-V9 (1934-1947)

SEND MS TO: Executive Secretary
                         Royal Statistical Society
                         25 Enford Street
                         London, W1H 2BH ENGLAND
                         Tel: 071-723-5882
                         (or) 071-723-6165
MS REQUIREMENT: Four copies, dbl spaced, summary of 100
words, 5-6 key words or phrases
PUBLISHER: Royal Statistical Society
DATE FOUNDED: 1948   FREQUENCY: 3/yr.
PRICE: $64.50   CIRC: 4500
LANGUAGE: English
INDEXED/ABSTRACTED: Appl Mech Rev, Compumath, Curr
Cont, Math R, SSCI
TARGET READER: Applied Statisticians, Applied Probabilists

Publishes papers on the theoretical and methodological aspects of statistics and its practice. Contributions included: new methods of collecting or analyzing data, or comparisons on new applications of existing methods, papers on the logical and philosophical basis of statistical theory, applied probability.

JOURNAL OF THE ROYAL STATISTICAL SOCIETY.
SERIES C: APPLIED STATISTICS
ISSN: 0035-9254

SEND MS TO: Executive Secretary
                         Royal Statistical Society
                         25 Enford Street
                         London, W1H 2BH ENGLAND
                         Tel: 071-723-5882
                         (or) 071-723-6165
MS REQUIREMENT: Five copies, dbl spaced, summary of 100
words, 5-6 key words or phrases

PUBLISHER: Royal Statistical Society
DATE FOUNDED: 1952   FREQUENCY: 3/yr.
PRICE: $64.50   CIRC: 5400
LANGUAGE: English
INDEXED/ABSTRACTED: Compumath, Math R, SSCI, Sci Abstr
TARGET READER: Applied Statisticians

Publishes papers motivated by real practical problems and aims to the
simple presentations of new or recent methodology. General Interest
Section accepts reviews of statistical practice in particular application
areas, surveys of the use of particular techniques, the eduational, and the
aspects of statistics.

JOURNAL OF THEORETICAL PROBABILITY
ISSN: 0894-9840

EDITOR: Arunava Mukherjea
                        Department of Math
                        University of South Florida
                        Tampa, FL 33620
MS REQUIREMENT: Original and two copies, dbl spaced, 100-
150 word abstract, 4-5 key words
PUBLISHER: Plenum
DATE FOUNDED: 1988   FREQUENCY: Quarterly
PRICE: (Inst) $150   LANGUAGE: English
                (Indv) $55   Back Issues Available
INDEXED/ABSTRACTED: Math R
TARGET READER: Probabilists

Publishes original papers in all areas of theoretical probability. Also
publishes authoritative expository papers and surveys of significant
emerging areas.

*JOURNAL OF TIME SERIES ANALYSIS**
ISSN: 0143-9782

PUBLISHER: Basil Blackwell
DATE FOUNDED: 1980   FREQUENCY: Quarterly
PRICE: $182   LANGUAGE: English
CIRC: 1,000   Back Issues Available
INDEXED/ABSTRACTED: Comput Control Abstr, Elect Electron
Abstr, Math R, Phys Abstr
TARGET READER: Time series analysts, Statisticians

Covers the basic theory and methodology of time series analysis.

JOURNAL OF UNDERGRADUATE MATHEMATICS
ISSN: 0022-5339

MANAGING   J. R. Boyd
EDITOR: Guilford College
                        Greensboro, NC 27410
                        Tel: (919) 292-5511
MS REQUIREMENT: Include the name of the institution where
the paper was written, the name of the faculty member who directed
the work and the class rank of the author at the time the paper was
completed.
PUBLISHER: Guilford College,
                        Department of Math
DATE FOUNDED: 1969   FREQUENCY: Bi-Annually
PRICE: (Inst) $9   LANGUAGE: English
                (Indv) $5
TARGET READER: Undergraduate Mathematics Students, College
Mathematics Teachers

Publishes research or expository papers written by undergraduate
students. A section, Proposal for Research, is a source of topics for
undergraduate research; solutions contributors to this section are both
faculty members and students.

K-THEORY
ISSN: 0920-3036

MANAGING   A. Bak
EDITOR: Department of Math
                   University of Bielefeld
                   Postfach 8640
                   4800 Bielefeld, GERMANY
MS REQUIREMENT: Triplicate, dbl spaced, no more than 150
words of abstract, short list of key words
PUBLISHER: Kluwer Academic Publishers
DATE FOUNDED: 1987   FREQUENCY: Quarterly
PRICE: $221.59   MICROFORM: UMI
LANGUAGE: English
INDEXED/ABSTRACTED: Math R
TARGET READER: Algebraists, K Theorists

Publishes papers on developments in the mathematical sciences which
are related to one of the various aspects of K-Theory. Original research
articles and research-survey articles are included. The latter provide
non-specialists with access and insight into topics of current research
and original works which alter existing accounts of the materials
surveyed by presenting new points of departure.

*KOBE JOURNAL OF MATHEMATICS**
ISSN: 0289-9051

HISTORY: Mathematics Seminar Notes
            [ISSN: 0385-633X] V1-V11 (1973-1983)

PUBLISHER: Kobe University
DATE FOUNDED: 1984   FREQUENCY: Semi-Annually
PRICE: Free   LANGUAGE: English
Order Direct
INDEXED/ABSTRACTED: Math R
TARGET READER: Mathematicians in all fields

Publishes research papers in all aspects of mathematics.

KUMAMOTO JOURNAL OF MATHEMATICS
ISSN: 0914-675X

HISTORY: *Kumamoto Journal of Science, Series A:*
              Mathematics, Physics, and Chemistry
              [ISSN: 0023-5318], V1-V8 (1952-1969)
              Kumamoto Journal of Science (Mathematics)
              [ISSN: 0385-6763], V9-V17 (1972-1987)

MANAGING   Masuyuki Hitsuda
EDITOR: Math Department
                  Faculty of Science
                  Kumamoto University
                  Kumamoto 860, JAPAN
MS REQUIREMENT: Duplicate
PUBLISHER: Kumamoto University
DATE FOUNDED: 1988   FREQUENCY: 2/yr.
PRICE: Order direct or   LANGUAGE: English
       on exchange basis
INDEXED/ABSTRACTED: Comput Control Abstr, Elect Electron
Abstr, Math R
TARGET READER: Mathematicians in all fields

Publishes mathematical research papers in all areas.

L'ENSEIGNEMENT MATHÉMATIQUE
ISSN: 0013-8584

EDITOR: R. Narasimhan
                  Department of Mathematics
                  University of Chicago
                  Chicago, IL 60637
PUBLISHER: L'Enseignement Mathématique
DATE FOUNDED: 1899   FREQUENCY: Quarterly
PRICE: $122.95   CIRC: 900
LANGUAGE: English, French, German, Italian
INDEXED/ABSTRACTED: Energy Res Abstr, Math R
TARGET READER: Mathematicians in all fields

Covers research-expository papers, survey and historical articles in mathematics.

LENINGRAD MATHEMATICAL JOURNAL
ISSN: 1048-9924

A cover to cover translation into English of papers published by the mathematics section of the Academy of Sciences of the USSR,

PUBLISHER: American Mathematical Society
DATE FOUNDED: 1990   FREQUENCY: Bi-Monthly
PRICE: (Inst) $785   LANGUAGE: English
              (Inst Memb) $628
INDEXED/ABSTRACTED: Math R, Zent Math
TARGET READER: Algebraists, Analysts

Publishes research papers, expository surveys and book reviews and contains contributions by some of the most prominent Soviet mathematical scientists.

LINEAR ALGEBRA AND ITS APPLICATIONS
ISSN: 0024-3795

EDITOR: Richard A. Brualdi
                         Department of Math
                         University of Wisconsin
                         Madison, WI 53706
MS REQUIREMENT: Duplicate, dbl spaced, up to 200 word
abstract
PUBLISHER: Elsevier
DATE FOUNDED: 1968   FREQUENCY: 51/yr.
PRICE: $1479   MICROFORM: RPI, UMI
LANGUAGE: English, French, German
INDEXED/ABSTRACTED: Compumath, Curr Cont, Eng Ind,
Math R, Sci Cit Ind
TARGET READER: Researchers who are interested in linear
algebra and matrix theory, researchers whose interests lie in

engineering, economics, computer science, operations research, and management science.

Publishes articles that contribute new information and new insights to matrix theory and finite dimensional linear algebra and their history in their algebraic, arithmetic, combinatorial, or numerical aspects, or that give significant applications to other branches of mathematics and other sciences.

LINEAR AND MULTILINEAR ALBEGRA
ISSN: 0308-1087

EDITOR: Marvin Marcus
Department of Computer Science
University of California
Santa Barbara, CA 93106

MS REQUIREMENT: Triplicate, dbl spaced, 100-150 word abstract, encourage manuscripts on computer disks, TEX, except for English, other languages require 200 word summary.

PUBLISHER: Gordon and Breach
DATE FOUNDED: 1973 FREQUENCY: Quarterly
PRICE: $1052 Language: English, French, German, Italian

INDEXED/ABSTRACTED: Int Abstr Oper Res, Math R
TARGET READER: Algebraists

Publishes research papers, research problems, expository or survey articles in linear and multilinear algebra and cognate areas.

LITHUANIAN MATHEMATICAL JOURNAL
English translation of Litovskii Matematicheskii Sbornik
ISSN: 0363-1672

HISTORY: *Mathematical Transactions of the Academy of Sciences of the Lithuanian SSR*
[ISSN: 0094-1719] V13 (1973)
Lithuanian Mathematical Transactions

[ISSN: 0148-8279] V14 (1974)

PUBLISHER: Consultants Bureau
DATE FOUNDED: 1975   FREQUENCY: Quarterly
PRICE: $545   LANGUAGE: English
INDEXED/ABSTRACTED: Math R, Zent Math
TARGET READER: Mathematicians in all fields

Focuses on fundamental problems on a wide variety of topics in
theoretical mathematics.

*MANUSCRIPTA MATHEMATICA**
ISSN: 0025-2611

PUBLISHER: Springer-Verlag
DATE FOUNDED: 1969   FREQUENCY: Monthly
PRICE: $823.64 LANGUAGE: English, French,
                                                               German
MICROFORM: UMI   REPRINT: ISI
INDEXED/ABSTRACTED: Compumath, Curr Cont, Ind Sci Rev,
Math R, Zent Math
TARGET READER: Mathematicians in all fields

Provides a forum for the rapid publication of advances in mathematical
research.

MATEKON
ISSN: 0025-1127

HISTORY: *Mathematical Studies in Economics and Statistics
                     in the USSR and Eastern Europe,*
                     V1-V5 (1964-1969)

EDITOR: John M. Litwack
                             Stanford University
PUBLISHER: Myron E. Sharpe
DATE FOUNDED: 1969   FREQUENCY: 4/yr.

PRICE: $287   MICROFORM: UMI
LANGUAGE: English
INDEXED/ABSTRACTED: CREJ, Compumath, Curr Cont, SSCI
TARGET READER: Mathematicians in all fields, Economists

Contains unabridged translations of articles from Soviet and East
European academic sources.

MATEMATICA APLICADA E COMPUTACIONAL
(Computational and Applied Mathematics)
ISSN: 0101-8205

EDITOR: C. S. Kubrusly
                          LNCC
                          R. Lauro Müller, 455
                          22,290 Rio de Janeiro, BRAZIL
                          FAX: (021) 295-8499
MS REQUIREMENT: Paper should be 15 pages long, brief
abstract, key words and phrases
PUBLISHER: Sociedade Brasileira de Matemática
                          Aplicada e Computacional
DATE FOUNDED: 1982   FREQUENCY: 3/yr.
PRICE: (Inst) $118.80   LANGUAGE: English, French,
                (Indv) $77.22   Portuguese
INDEXED/ABSTRACTED: Curr Cont, Math R, Zent Math
TARGET READER: Applied Mathematicians, Mathematically
oriented Engineers, and Computer Scientists

Publishes original works in any area of applied mathematics. It features
special issues on scientific computing, devoted to those numerical,
non-numerical and statistical techniques designed to solve scientific and
technological problems with the aid of computers.

MATHEMATICA JAPONICA
ISSN: 0025-5513

MANAGING   T. Ishihara
EDITOR: Japanese Assoc. of Mathematical Sciences
                      Shin Sakaihigashi Bldg.
                      2-1-18 Minami Hanadaguchi
                      Sakai, Osaka 590, JAPAN
                      Tel: 0722-22-1850
                      FAX: 0722-22-7987
MS REQUIREMENTS: Original and two copies, dbl spaced,
abstract less than 150 words, 1980 AMS Subject Classification
Scheme, accept AMS-TEX form
PUBLISHER: Japanese Assoc. of Mathematical Sciences
DATE FOUNDED: 1948   FREQUENCY: Bi-Monthly
PRICE: $330   CIRC: 850
LANGUAGE: English, French, German
PUBLICATION CHARGES: $40/Page
INDEXED/ABSTRACTED: Math R
TARGET READER: Mathematicians, Statisticians, Computer
Scientists, Operations Researchers

Publishes original papers submitted from all over the world. Devoted
to contributions in the fields of pure mathematics, statistics, computer
science, operations research, applied mathematics and other
mathematical sciences.

MATHEMATICA SCANDINAVICA
ISSN: 0025-5521

HISTORY: *Tidsskrift for Mathematik* (1859-1889)
                  Nyt Tidsskrift for Matematik (1890-1919)
                  Matematisk Tidsskrift. B (1919-1952)

EDITORICAL   Mathematica Scandinavica
OFFICE: Matematisk Institut
                  NY Munkegade, 8000 Arhus C
                  DENMARK
MS REQUIREMENT: Duplicate, sufficient spacing
PUBLISHER: Dansk Matematisk Forening/
                  Matematisk Institut

DATE FOUNDED: 1953   FREQUENCY: Quarterly
PRICE: $161.81   CIRC: 1100
Back Issues Available   LANGUAGE: English, French,
                                     German
INDEXED/ABSTRACTED: Compumath, Ind Sci Rev, Math R
TARGET READER: Mathematicians in all fields

Publishes mathematical papers of moderate length, especially by Scandinavian authors.

MATHEMATICA SLOVACA
ISSN: 0139-9918

HISTORY: *Matematicko-fyzikbalny Sbornik Slovenskej*
               akadbemie vied a umenbi (1951-1952)
               Matematicko-fyzikbalny Icasopis
               [ISSN: 0543-0046] (1953-1966)
               Matematicky Icasopis [ISSN: 0025-5173]
               (1967-1975)

PUBLISHER: Slovenska Academia Vied
DATE FOUNDED: 1976   FREQUENCY: Quarterly
PRICE: $108.39   CIRC: 1,000
LANGUAGE: English, French, German, Russian
INDEXED/ABSTRACTED: Math R
TARGET READER: Mathematicians in all fields

Publishes original scientific papers of Czechoslovak as well as foreign authors from various mathematical disciplines.

MATHEMATICAL AND COMPUTER MODELLING
ISSN: 0895-7177

HISTORY: *Mathematical Modelling*
               [ISSN: 0270-0255] V1-V9 (1980-1987)

EDITOR: Ervin Y. Rodin

Department of Systems Science and Math
Campus Box 1040
Washington University
1 Brookings Dr.
St. Louis, MO 63130

MS REQUIREMENT: Triplicate, dbl spaced, short abstract, six
possible reviewers' names and full addresses, AMS-TEX form
accepted.

PUBLISHER: Pergamon Press
DATE FOUNDED: 1988   FREQUENCY: Monthly
PRICE: $635   MICROFORM: UMI,MIM,RPI
LANGUAGE: English   Back Issues from Pergamon
INDEXED/ABSTRACTED: Compumath, Comput Control Abstr,
Curr Cont, Elect Electron Abstr, Math R, Phys Abstr, Sci Abstr
TARGET READER: Mathematicians, Computer Scientists,
Physicists

Publishes both theoretical and applied papers to provide a medium of
exchange for the diverse disciplines utilizing mathematical or computer
modelling as either a theoretical or working tool. Equal attention will
be given to the mechanics, methodology, and theory of modelling with
an attempt to advocate either mathematical or computer modelling, or
a combination of the two in an interactive form.

MATHEMATICAL BIOSCIENCES
ISSN: 0025-5564

EDITOR: John A. Jacquez
                       Department of Physiology
                       University of Michigan
                       7808 Medical Sciences II
                       Ann Arbor, MI 48109-0622

MS REQUIREMENT: Duplicate, dbl spaced, up to 200 word
abstract

PUBLISHER: Elsevier
DATE FOUNDED: 1967   FREQUENCY: 10/yr.
PRICE: $675   MICROFORM: RPI, UMI

LANGUAGE: English   CIRC: 1000
INDEXED/ABSTRACTED: Appl Mech Rev, Biol Abstr, Chem
Abstr, Curr Cont, Excerpta Med, Eng Ind, Math R
TARGET READER: Biomathematicians, Statisticians,
Epidemiologists

Publishes research and expository mathematical papers devoted to the
formulation, analysis, and numerical solution of mathematical models
in the biosciences.

MATHEMATICAL CHRONICLE
ISSN: 0581-1155

EDITOR: S. Fitzpatrick
MS REQUIREMENT: Duplicate, AMS-TEX form accepted
PUBLISHER: University of Auckland
                        Mathematical Chronicle Committee
                        Department of Math and Statistics
DATE FOUNDED: 1969   FREQUENCY: Irr
PRICE: $14   CIRC: 300
LANGUAGE: English
INDEXED/ABSTRACTED: Math R
TARGET READER: Mathematicians in all fields.

Publishes both long and short expository and survey articles on any area
of mathematics. Short research articles and mathematical notes limited
to 5 printed pages. Also includes mathematical education notes,
problems and solutions, book reviews, news items and announcements,
personal notes.

MATHEMATICAL FINANCE
ISSN: 0960-1627

EDITOR: Stanley Pliska
PUBLISHER: Basil Blackwell
DATE FOUNDED: 1991   FREQUENCY: Quarterly
PRICE: $132   LANGUAGE: English

MICROFORM: UMI
TARGET READER: Business Managers

Publishes articles to serve as forum for financial researchers, financial practitioners and mathematical scientists.

MATHEMATICAL GAZETTE*
ISSN: 0025-5572

HISTORY: *Report of the Association for the Improvement of Geometrical Teaching* (1871-1894)

PUBLISHER: Mathematical Association
DATE FOUNDED: 1894   FREQUENCY: Bi-Monthly
PRICE: $55.82   LANGUAGE: English
CIRC: 5,000   MICROFORM: UMI
REPRINT: UMI
INDEXED/ABSTRACTED: Ind Sci Rev, Math R
TARGET READER: Mathematics educators

Publishes papers on mathematical topics of wide appeal and mathematics teaching at the level of secondary school, college, and university.

MATHEMATICAL GEOLOGY*
ISSN: 0882-8121

HISTORY: *Journal of the International Association for Mathematical Geology* [ISSN: 0020-5958] V1-V17 (1969-1985)

PUBLISHER: Plenum
DATE FOUNDED: 1986   FREQUENCY: Monthly
PRICE: $360   LANGUAGE: English
Back Issues Available
INDEXED/ABSTRACTED: Appl Mech Rev, Chem Abstr, Compumath, Curr Cont, Eng Ind, Math R, Zent Math

TARGET READER: Geologists

Publishes original papers which are of broad interest to the discipline of geology.

MATHEMATICAL INTELLIGENCER
ISSN: 0343-6993

SEND MS TO:	Chandler Davis
	Department of Math
	University of Toronto
	Toronto, CANADA, M5S 1A1
MS REQUIREMENT:	Duplicate, dbl spaced
PUBLISHER:	Springer-Verlag

DATE FOUNDED: 1978   FREQUENCY: Quarterly
PRICE: $35   LANGUAGE: English
MICROFORM: UMI   REPRINT: ISI
INDEXED/ABSTRACTED: Gen Sci Ind, Ind Sci Rev, Math R, Zent Math
TARGET READER: Mathematicians in all fields

Publishes articles about mathematics, mathematicians, history and culture of mathematics. Articles inform and entertain a broad audience of mathematicians.

*MATHEMATICAL JOURNAL OF OKAYAMA UNIVERSITY**
ISSN: 0030-1566

PUBLISHER: Okayama University
DATE FOUNDED: 1952   FREQUENCY: Semi-Annually
PRICE: Exchange Only   CIRC: 700
LANGUAGE: English, French, German
INDEXED/ABSTRACTED: Math R, Zent Math
TARGET READER: Mathematicians in all fields

Publishes research papers in mathematics.

MATHEMATICAL METHODS IN THE APPLIED SCIENCES
ISSN: 0170-4214

EDITOR: B. Brosowski
                             Fachbereich Mathematik
                             der Universitat, GERMANY
MS REQUIREMENT: Duplicate, dbl spaced, MOS Classification
Numbers, 100 word summary
PUBLISHER: B. G. Teubner
DIST BY: John Wiley
DATE FOUNDED: 1979   FREQUENCY: 9/yr.
PRICE: $450   CIRC: 1000
LANGUAGE: English
INDEXED/ABSTRACTED: Appl Mech Rev, Compumath, Curr
Cont, Eng Ind, Math R, Zent Math
TARGET READER: Applied Mathematicians, Educators in applied
mathematics and applied sciences

Publishes papers which clearly demonstrate the development and
constructive application of mathematics to significant problems in
science.

MATHEMATICAL MODELS AND METHODS IN APPLIED SCIENCES
ISSN: 0218-2025

EDITOR: Nicola Bellomo
                     Dipartimento di Matematica
                     Politecnico di Torino
                     Corso Duca degli Abruzzi, 24
                     1-10129 Torino, ITALY
                     FAX: 39-11-5647599
MS REQUIREMENT: Triplicate, dbl spaced, abstract not to
exceed 300 words
PUBLISHER: World Scientific Publishing Co.
DATE FOUNDED: 1991   FREQUENCY: Quarterly
PRICE: $190   LANGUAGE: English

TARGET READER: Applied Mathematicians

Publishes papers on mathematical modelling dynamic systems, partial differential equations.

MATHEMATICAL NOTES OF THE ACADEMY OF SCIENCE OF THE USSR
ISSN: 0001-4346

Translation of *Matematicheskie Zametki,* a publication of the Academy of Sciences of the USSR.

PUBLISHER: Consultants Bureau
DATE FOUNDED: 1952   FREQUENCY: Monthly
PRICE: $915   LANGUAGE: English
INDEXED/ABSTRACTED: Appl Mech Rev, Curr Cont, Math R, Zent Math
TARGET READER: Mathematicians in all fields

Covers original papers in all branches of contemporary mathematics.

MATHEMATICAL POPULATION STUDIES
ISSN: 0889-8480

COORDINATING   Marc Artzrouni
EDITOR: Department of Mathematical Sciences
                              Loyola University
                              6363 St. Charles Avenue
                              New Orleans, LA 70118
MS REQUIREMENT: Triplicate, dbl spaced, 150 word abstract,
six key words
PUBLISHER: Gordon and Breach
DATE FOUNDED: 1988   FREQUENCY: 4/yr.
PRICE: Varies   LANGUAGE: English
INDEXED/ABSTRACTED: Math R
TARGET READER: Mathematicians, Statisticians, Sociologists,

Economists, Biologists

Publishes research papers in the mathematical and statistical study of human populations. Serves as a forum for the exchange of views between researchers in academia, international organizations, research institutes, and statistical offices throughout the world.

MATHEMATICAL PROCEEDINGS OF THE CAMBRIDGE PHILOSOPHICAL SOCIETY
ISSN: 0305-0041

HISTORY: *Proceedings of the Cambridge Philosophical Society--Mathematical and Physical Sciences*
[ISSN: 0008-1981] V1-V76 (1843-1974)

SEND MS TO: Editor, Mathematical Proceedings
Cambridge Philosophical Society
Bene't St.
Cambridge CB2 3PY ENGLAND
MS REQUIREMENT: Duplicate, dbl spaced, brief abstract
PUBLISHER: Cambridge University Press
DATE FOUNDED: 1975   FREQUENCY: Bi-Monthly
PRICE: $299   LANGUAGE: English
MICROFORM: UMI   Back Issues Available
INDEXED/ABSTRACTED: Appl Mech Rev, Chem Abstr, Compumath, Curr Cont, Math R, Sci Abstr, Sci Cit Ind, Zent Math
TARGET READER: Mathematicians in all fields

Publishes papers which advance knowledge of mathematics, either pure or applied.

MATHEMATICAL PROGRAMMING (Series A & B)
ISSN: 0025-5610

HISTORY: *Mathematical Programming*
[ISSN: 0025-5610] V1-V39 (1971-1987)

EDITOR: Ŕ. Bixby
                               Department of Mathematical Sciences
                               Rice University
                               Houston, TX 77251
MS REQUIREMENT: Series A/Four copies, abstract not to
exceed 200 words, key word, abbreviated title for running head.
PUBLISHER: North-Holland
DATE FOUNDED: 1988   FREQUENCY: 12/yr.
PRICE: $600   MICROFORM: RPI
LANGUAGE: Any language/English is official language
INDEXED/ABSTRACTED: ACM Guide Comput Lit, Eng Ind, ISI
Curr Cont, Math R
TARGET READER: Mathematicians, Systems Engineers, Electrical
Engineers, Economists, Operations Researchers

Publishes original work dealing with every theoretical, computational
and applicational aspect of mathematical programming.   Series A
publishes original research articles, expositions and surveys, and reports
on computational experimentation and new or innovative practical
applications as well as short communications dealing with the above.
Series B focuses on a single subject, selected to respond to the current
interests of the mathematical programming community and has one or
more guest-editors. An issue may be a collection of original articles,
a single research monograph, or a selection of papers from an
appropriate conference.

MATHEMATICAL REPORTS
ISSN: 0275-7214

EDITOR: Prof. G. F. D. Duff
                               Department of Math
                               University of Toronto
                               Toronto, Ontario
                               M5S 1A1 CANADA
MS REQUIREMENT: Typewritten, dbl spaced, short abstract
PUBLISHER: University of Toronto
                               Department of Math

DATE FOUNDED: 1983   FREQUENCY: Irr
PRICE: $15   LANGUAGE: English, French
PUBLICATION CHARGES: $15/page
INDEXED/ABSTRACTED: Math R
TARGET READER: Mathematicians in all fields

Publishes short papers which are limited to six typed pages
summarizing completed research.

*MATHEMATICAL REPORTS OF THE ACADEMY OF SCIENCE**
ISSN: 0706-1994

EDITORIAL   Mathematical Reports
OFFICE: Department of Pure Mathematics
                        University of Waterloo
                        Waterloo, Ontario, N2L 3G1 CANADA
PUBLISHER: Royal Society of Canada
DATE FOUNDED: 1979   FREQUENCY: Irr
PRICE: Varies   LANGUAGE: English, French
INDEXED/ABSTRACTED: Math R
TARGET READER: Mathematicians in all fields

Publishes original papers in both pure and applied mathematics.

MATHEMATICAL REVIEWS
ISSN: 0025-5629

EDITOR: G. J. Janusz (Edited at the
                        University of Michigan)
                        P. O. Box 8604
                        416 Fourth Street
                        Ann Arbor, MI 48107-8604
PUBLISHER: American Mathematical Society
DATE FOUNDED: 1940   FREQUENCY: Monthly
PRICE: (Inst) $4010   MICROFORM: AMS, UMI
              (Inst Memb) $3208   CIRC: 2400

(Indv Memb) $481   LANGUAGE: English, French,
(Reviewer) $321   German, Italian
INDEXED/ABSTRACTED: Appl Mech Rev, Math R
TARGET READER: Mathematicians in all fields.

This publication is recognized as a reviewing and abstracting journal
covering published mathematical research literature. Reviewers are
assigned from among 14,000 mathematicians around the world. Over
45,000 reviews or abstracts are published each year in twelve monthly
issues. Items are arranged in each issue according to the 1980
Mathematics Subject Classification (1985 Revision). Annual author and
subject indexes are included with the subscription. Cumulative author
and/or subject indexes are available. Subscriptions are available on
paper and microfiche.

MATHEMATICAL SCIENTIST
ISSN: 0312-3685

EDITOR: Joseph M. Gani
                           Department of Statistics and Applied
                           Probability
                           University of California
                           Santa Barbara, CA 93106
MS REQUIREMENT: Dbl spaced, short abstract, list of key
words
PUBLISHER: Applied Probability Trust
DATE FOUNDED: 1976   FREQUENCY: Semi-Annually
PRICE: $13.50   LANGUAGE: English, French
CIRC: 750
INDEXED/ABSTRACTED: Curr Ind Stat, J. Cont Quant Meth,
Math R
TARGET READER: Probabilists, Statisticians, Mathematicians in all
fields

Publishes research papers of general interest, and uses mathematical
theory, methods and models to provide insight into the phenomena
studies in the engineering, physical, biological and social sciences.

MATHEMATICAL SOCIAL SCIENCE
ISSN: 0165-4896

EDITOR:	Ki Hang Kim
	Mathematics Research Group
	Alabama State University
	P. O. Box 271
	Montgomery, AL 36101-0271
MS REQUIREMENT:	Triplicate, dbl spaced, abstract no more
than 200 words	
PUBLISHER:	North-Holland
DATE FOUNDED:	1980 FREQUENCY: Bi-Monthly
PRICE:	$368.53 MICROFORM: RPI, UMI
LANGUAGE:	English

INDEXED/ABSTRACTED: Cyb Abstr, Math R, SSCI, Sci Abstr, Socio Abstr, Zent Math
TARGET READER: Economists, Social Scientists using Quantitative Methods, Mathematicians, Operations Researchers, Statisticians

Publishes original research papers, survey papers, short notes, news items, a calendar of meetings, and book reviews which are of broad interest in the mathematical social sciences.

MATHEMATICAL SPECTRUM
ISSN: 0025-5653

MANAGING	Joseph M. Gani
EDITOR:	Department of Statistics and
	Applied Probability
	University of California
	Santa Barbara, CA 93106
SEND MS TO:	The Editor, Mathematical Spectrum
	Hicks Bldg, The University
	Sheffield S3 7RH, ENGLAND
PUBLISHER:	Applied Probability Trust
DATE FOUNDED:	1968 FREQUENCY: 4/yr.
PRICE:	$13 CIRC: 2000

LANGUAGE: English
INDEXED/ABSTRACTED: Math R, Ref Zh, Sci Abstr
TARGET READER: Students, Teachers in schools, colleges and
universities, General Readers interested in mathematics.

Publishes articles that deal with the entire range of mathematical
discipline, pure and applied mathematics, statistics, operational research,
computer science, numerical analysis and biomathematics. Expository,
historical material, elementary research, information or educational
opportunities and careers in mathematics all are included.

MATHEMATICAL SYSTEMS THEORY
ISSN: 0025-5661

EDITOR: Arnold L. Rosenberg, Computer and
                        Information Sciences
                        University of Massachusetts
                        Amherst, MA 01003
PUBLISHER: Springer-Verlag
DATE FOUNDED: 1966   FREQUENCY: 4/yr.
PRICE: $106   MIRCOFORM: UMI
LANGUAGE: English   REPRINTS: ISI
INDEXED/ABSTRACTED: Compumath, Curr Cont, Ind Sci Cit,
Math R, Sci Abstr, Zent Math
TARGET READER: Computer Scientists, Economists,
Biologists, Applied Mathematicians

Publishes research papers in the theories of both discrete and continuous
systems. Examines the various mathematical aspects of everyday
problems in engineering, computer science, economics, and biology. It
covers theory of algorithms and computational complexity, mathematical
aspects of programs, languages, automata, and computing systems, etc.

MATHEMATICS AND COMPUTER EDUCATION
ISSN: 0730-8639

HISTORY: *MATYC Journal* [ISSN: 0092-1424] V1-V15
                (1967-1981)

SEND MS TO: Mathematics and Computer Education
                      P. O. Box 158
                      Old Bethpage, NY 11804
MS REQUIREMENT: Six copies, dbl spaced
PUBLISHER: MATYC Journal, Inc.
DATE FOUNDED: 1982   FREQUENCY: 3/yr.
PRICE: (Inst) $50   LANGUAGE: English
             (Indv) $23   CIRC: 3000
MICROFORM: UMI   REPRINT: UMI
INDEXED/ABSTRACTED: CIJE, Comput Rev, Math R, Sci Abstr
TARGET READER: Mathematics Educators, Undergraduate
Students

Publishes articles devoted to critical evaluation and dissemination of
articles, development of materials for the improvement of classroom
effectiveness in the first years of college as well as senior high schools,
encouragement of high academic standards.

MATHEMATICS AND COMPUTERS IN SIMULATION
ISSN: 0378-4754

HISTORY: *Annales de l'Association Internationale pour le
                Calcul Analogique* [ISSN: 0020-594X]
                V1-V17 (1958-1975)
                *Transactions of the International Association for
                Mathematics and Computers in Simulation*
                [ISSN: 0377-9114] V18 (1976)

EDITOR: R. Vichnevetsky
                      Department of Computer Science
                      Rutgers University
                      New Brunswick, NJ 08903
MS REQUIREMENT: Triplicate, dbl spaced, left margin at least
5cm, camera ready copy is accepted (single space) summary of 10 to

20 typed lines
PUBLISHER: North-Holland
DATE FOUNDED: 1977   FREQUENCY: Bi-Monthly
PRICE: $247.19   MICROFORM: RPI, UMI
CIRC: 1000   LANGUAGE: English, French
INDEXED/ABSTRACTED: ACM Guide Comput Lit, Cam Sci
Abstr, Comput Abstr, Eng Ind, Math R
TARGET READER: Applied Mathematicians, Statisticians,
System Analysts, Simulation Specialists

Provides an international forum for the dissemination of up-to-date
information in the field of the computer simulation of systems. Short
and concise research papers and general tutorial articles are included.

MATHEMATICS AND ITS APPLICATIONS*
ISSN: 0543-0941

HISTORY: *Notes on Mathematics and Its Application*
                [ISSN: 0888-6113] (1961-1970)

PUBLISHER: Gordon and Breach
DATE FOUNDED: 1971   FREQUENCY: Irr
PRICE: $215   LANGUAGE: English
INDEXED/ABSTRACTED: Math R
TARGET READER: Mathematicians in all fields

Publishes papers in all areas of mathematics.

MATHEMATICS JOURNAL OF TOYAMA UNIVERSITY
ISSN: 0916-6009

HISTORY: *Mathematics Reports, Toyama University*
                [ISSN: 0386-832X] V1-V12 (1978-1989)

MS REQUIREMENT: Typewritten, 1½ line spaced, abstract
optional

PUBLISHER: Toyama University
                        Department of Mathematics
                        Faculty of Science
DATE FOUNDED: 1990   FREQUENCY: Irr
PRICE: Request from   LANGUAGE: English
                Publisher
INDEXED/ABSTRACTED: Elect Electron Abstr, Math R, Phys
Abstr
TARGET READER: Mathematicians in all fields

Publishes research papers in pure and applied mathematics.

MATHEMATICS MAGAZINE
ISSN: 0025-570X

HISTORY: *Mathematics News Letters,*
                   V1-V8 (1926-1934)
                National Mathematics Magazine,
                   V9-V20 (1934-1945)

EDITOR: Martha J. Siegel
                        Towson State University
                        Towson, MD 21204
MS REQUIREMENT: Original plus two copies, dbl spaced, 1980
AMS Subject Classification Numbers
PUBLISHER: Mathematical Assoc. of America
DATE FOUNDED: 1947   FREQUENCY: Bi-Monthly
PRICE: (Inst) $64   Except July/August
               (Indv) $16   MICROFORM: UMI
CIRC: 14,000   REPRINTS: UMI
LANGUAGE: English
INDEXED/ABSTRACTED: Gen Sci Ind, Ind Sci Rev, Math R
TARGET READER: Undergraduate and Graduate students of
mathematics, Mathematics Teachers

Presents papers of lively and appealing mathematical expositions.
Research papers are not published.   Articles include examples,
applications, historical background, and illustrations which should

be attractive and accessible to undergraduates, be helpful in supplementing undergraduate courses and in stimulating student investigations.

MATHEMATICS OF COMPUTATION
ISSN: 0025-5718

EDITOR: Walter Gautschi
                                 Department of Computer Science
                                 Purdue Univeristy
                                 West Lafayette, IN   47907
PUBLISHER: American Mathematical Society
DATE FOUNDED: 1943   FREQUENCY: Quarterly
PRICE: (Inst) $205   CIRC: 2500
          (Inst Memb) $164   MICROFORM: UMI, AMS
          (Indv Memb) $123   LANGUAGE: English
INDEXED/ABSTRACTED: Appl Mech Rev, Compumath, Cyb
Abstr, Math R, Sci Abstr
TARGET READER: Mathematicians interested in computations,
Computer Scientists

Publishes original articles on all aspects of numerical mathematics, book reviews, mathematical talks, and technical notes. Devoted to advances in numerical analysis, the application of computational methods, high speed calculating and other aids to computation.

MATHEMATICS OF CONTROL, SIGNALS, AND SYSTEMS
ISSN: 0932-4194

EDITOR: Eduardo D. Sontag
                                 Department of Math
                                 Rutgers University
                                 New Brunswick, NJ 08903
MS REQUIREMENT: Five copies, no longer than 50 pages, dbl
spaced, detailed abstract, 5-10 key words, accept LATEX form.
PUBLISHER: Springer Verlag

DATE FOUNDED: 1988   FREQUENCY: Quarterly
PRICE: $188   LANGUAGE: English
MICROFORM: UMI
INDEXED/ABSTRACTED: Comput Control Abstr, Elect
Electron Abstr, Math R
TARGET READER: Mathematicians, Engineers

Publishes original research papers in the areas of mathematical systems
theory, control theory, and signal processing.

MATHEMATICS OF OPERATIONS RESEARCH
ISSN: 0364-765X

EDITOR: Erhan Cinlar
                     Department of Civil Engineering and
                     Operations Research
                     Princeton University
                     Princeton, NJ 08544
MS REQUIREMENT: Four copies, dbl spaced, abstract up to
200 words, key words
PUBLISHER: Institute of Management Sciences
DATE FOUNDED: 1976   FREQUENCY: Quarterly
PRICE: (Inst) $68   LANGUAGE: English
              (Indv) $35
INDEXED/ABSTRACTED: Compumath, Cyb Abstr, Ind Sci Rev,
Math R, Sci Abstr
TARGET READER: Mathematicians, Operations Researchers,
Management Scientists

Publishes significant research and review papers having substantial
mathematical interest and relevance to operations research and
management science.

MATHEMATICS OF THE USSR-IZVESTIYA
ISSN: 0025-5726

English translation of: *Izvestiya Matematicheskaja Serija*

TRANSLATION   Ben Silver
EDITOR: 77-E Charles Street
                         Providence, RI 02904
PUBLISHER: American Mathematical Society
DATE FOUNDED: 1967   FREQUENCY: Bi-Monthly
PRICE: (Inst) $661   MICROFORM: AMS, UMI
                (Indv) $529   CIRC: 700
LANGUAGE: English
INDEXED/ABSTRACTED: Compumath, Curr Cont, Ind Sci Rev,
Math R
TARGET READER: Mathematicians

A cover to cover translation into English of Izvestiya Akademii Nauk
SSR Seriya Matematicheskaya, a journal of current areas of pure
mathematics research, which is published by the Academy of Sciences
of the USSR.

MATHEMATICS OF THE USSR-SBORNIK
ISSN: 0025-5734

English translation of: *Matematicheskii Sbornik*

TRANSLATION   Ben Silver
EDITOR: 77-E Charles Street
                         Providence, RI 02904
PUBLISHER: American Mathematical Society
DATE FOUNDED: 1967   FREQUENCY: Bi-Monthly
PRICE: (Inst) $801   CIRC: 700
                (Indv) $641   LANGUAGE: English
MICROFORM: AMS, UMI
INDEXED/ABSTRACTED: Compumath, Curr Cont, Ind Sci Rev,
Math R
TARGET READER: Mathematicians in all fields

A cover-to-cover translation into English of Matematicheskii Sbornik,
published by the Moscow Mathematical Society and the Academy of
Sciences of the USSR.

MATHEMATICS TEACHER
ISSN: 0025-5769

EDITOR: Jerry Johnson
Western Washington University
Bellingham, WA 98225
PUBLISHER: National Council of Teachers
of Mathematics
DATE FOUNDED: 1908 FREQUENCY: Monthly
PRICE: (Inst) $45 Except June, July, August
(Indv) $40 MICROFORM: UMI, MIM
CIRC: 44200 LANGUAGE: English
INDEXED/ABSTRACTED: Biog Ind, Educ Ind, Except Child Educ
Abstr, Math R
TARGET READER: Junior and Senior High School Teachers,
Mathematics Instructors in Junior Colleges and Educators.

Devoted to the improvement of the mathematics instruction in Junior
High Schools, Senior High Schools, Two-Year Colleges and
Teacher-Education Colleges.

MATHEMATIKA
ISSN: 0025-5793

EDITOR: P. McMullen
Department of Math
University College, London
Gower Street
London, WC1E 6BT, ENGLAND
PUBLISHER: University College, London
Department of Math
ORDERS TO: Mathematika
J. W. Arrowsmith, Ltd.
DATE FOUNDED: 1954 FREQUENCY: Semi-Annually
PRICE: $81.20 LANGUAGE: English
CIRC: 700 Back Issues Available
INDEXED/ABSTRACTED: Appl Mech Rev, Compumath, Curr
Cont, Ind Sci Rev, Math R, Zent Math

TARGET READER: Mathematicians in all fields

Contains original notes and memoirs on mathematics and its applications.

MATHEMATISCHE ANNALEN
ISSN: 0025-5831

SEND MS TO: Individual Subject Editor
                        not the Managing Editor
PUBLISHER: Springer Verlag
DATE FOUNDED: 1868   FREQUENCY: 12/yr.
PRICE: $1726.79 MICROFORM: UMI
REPRINT: ISI   LANGUAGE: English, French,
                                            German, Italian
INDEXED/ABSTRACTED: Compumath, Curr Cont, Ind Sci Rev,
Math R, Zent Math
TARGET READER: Mathematicians interested in complex analysis,
algebraic geometry, algebraic number theory, functional analysis,
differential geometry.

Publishes papers in a broad range of mathematics. Although basically
unspecialized, the journal covers especially the field of analysis.

MATHEMATISCHE NACHRICHTEN
ISSN: 0025-584X

EDITOR: Helmut Koch
                        Redaktion Mathematische Nachrichten
                        Karl-Weierstrass-Institute
                          für Mathematik
                        Mohrenstrasse 39
                        0-1086 Berlin, GERMANY
PUBLISHER: Akademie Verlag
DATE FOUNDED: 1948   FREQUENCY: Irr

PRICE: Varies   LANGUAGE: English, French,
                                                   German, Russian
INDEXED/ABSTRACTED: Compumath, Curr Cont, Ind Sci Rev,
Math R
TARGET READER: Mathematicians in all fields

Publishes original mathematical papers.

MATHEMATISCHE ZEITSCHRIFT
ISSN: 0025-5874

EDITOR: Manuscripts sent to
                          various subject editors
MS REQUIREMENT: Duplicate, dbl spaced, summary including
key words for classification
PUBLISHER: Springer-Verlag
DATE FOUNDED: 1918   FREQUENCY: Monthly
PRICE: $1527.54 LANGUAGE: English
MICROFORM: UMI   REPRINT: ISI
INDEXED/ABSTRACTED: Compumath, Curr Cont, Ind Sci Rev,
Math R, Zent Math
TARGET READER: Mathematicians in both pure and applied fields,
Physicists, Astronomers.

Devoted to pure and applied mathematics, but papers on theoretical
physics and astronomy will be accepted if they present interesting
mathematical results.

MEMOIRS OF THE AMERICAN MATHEMATICAL SOCIETY
ISSN: 0065-9266

PUBLISHER: American Mathematical Society
DATE FOUNDED: 1950   FREQUENCY: Bi-Monthly
PRICE: (Inst) $270   CIRC: 1200
        (Inst Memb) $216   LANGUAGE: English
INDEXED/ABSTRACTED: Compumath, Math R, Zent Math

TARGET READER: Mathematicians in all fields.

Devoted to long research papers in pure and applied mathematics. Each paper (in some cases a group of shorter cognate papers) is published as an individual number and may be ordered separately.

METRIKA*
ISSN: 0026-1335

HISTORY: *Mitteilungsblatt für Mathematische Statistik*
              [ISSN: 0176-6074] Bd1-Bd6 (1949-1954)
      merged with
              Statistische Vierteljarhsschrift
              ISSN: 0255-7479) Bd1-Bd10 (1948-1957)

PUBLISHER: Springer-Verlag
DATE FOUNDED: 1958   FREQUENCY: Bi-Monthly
PRICE: $223.33 LANGUAGE: English
CIRC: 1,000
INDEXED/ABSTRACTED: Curr Cont, Int Abstr Oper Res, J. Cont
Quant Meth, Math R, Sci Abstr
TARGET READER: Theoretical and Applied Statisticians

Covers articles on statistical methods and mathematical statistics.

MICHIGAN MATHEMATICAL JOURNAL
ISSN: 0026-2285

EDITOR: Douglas G. Dickson
                        Math Department
                        University of Michigan
                        Ann Arbor, MI 48109
SEND MS TO: The Editor
                        Michigan Mathematical Journal
                        University of Michigan
                        Ann Arbor, MI 48109
MS REQUIREMENT: Duplicate

PUBLISHER: University of Michigan
DATE FOUNDED: 1953 FREQUENCY: 3/yr.
PRICE: (Inst) $50 CIRC: 900
(Indv) $25 LANGUAGE: English, French
PUBLICATION CHARGE: $25/page to the author's institution
Back Volumes Available: University of Michigan
and Johnson Reprint Corp.
INDEXED/ABSTRACTED: Compumath, Curr Cont, Ind Sci Rev,
Math R
TARGET READER: Mathematicians in all fields

Publishes research articles in all areas of mathematics.

MONATSHEFTE FÜR MATHEMATIK
ISSN: 0026-9255

HISTORY: *Monatshefte für Mathematik und Physik,*
V1-V51 (1890-1944)

SEND MS TO: Monatshefte für Mathematik
Institut für Mathematik der Universität
Strudlhofgasse 4,
A-1090 Wien, AUSTRIA
MS REQUIREMENT: Duplicate, dbl spaced, English summary,
papers in French and German need English translation of title
PUBLISHER: Springer-Verlag
DATE FOUNDED: 1948 FREQUENCY: Monthly
PRICE: $459 MICROFORM: UMI
REPRINT: ISI LANGUAGE: English,
German, French
INDEXED/ABSTRACTED: Compumath, Curr Cont, Ind Sci Rev,
Math R, Zent Math
TARGET READER: Mathematicians in the whole field of pure
mathematics

Publishes research papers in pure mathematics in its broadest
significance. Topics cover Algebra, Real and Complex Analysis,

Biomathematics, Dynamical Systems, Functional Analysis, Number Theory, etc.

MOSCOW UNIVERSITY MATHEMATICS BULLETIN
ISSN: 0027-1322

English translation of *Vestnik Moskovskogo Universiteta, Matenatika*

EDITOR:	O. B. Lupanor		
PUBLISHER:	Allerton Press		
DATE FOUNDED:	1986	FREQUENCY:	Bi-Monthly
PRICE:	$640	LANGUAGE:	English

INDEXED/ABSTRACTED: Appl Mech Rev, Math R
TARGET READER: Mathematicians in all fields

Covers both pure and applied mathematical articles.

NAGOYA MATHEMATICAL JOURNAL
ISSN: 0027-7630

SEND MS TO: Managing Editor
                    Nagoya Mathematical Journal
                    Department of Math/School of Science
                    Nagoya University, Chikusa-ku
                    Nagoya 464-01, JAPAN
MS REQUIREMENT: Sufficient space between lines, follow standard style usual for mathematical literature
PUBLISHER: Univeristy of Nagoya
SUBSCRIPTION
IN U.S. TO: Kinokuniya Co., Ltd.
DATE FOUNDED: 1950 FREQUENCY: Quarterly
PRICE: $230 LANGUAGE: English, French, German

INDEXED/ABSTRACTED: Compumath, Curr Cont, Ind Sci Rev, Math R, Sci Cit Ind

TARGET READER: Mathematicians in all fields

Primarily publishes research papers, also features invited papers on mathematics.

NAVAL RESEARCH LOGISTICS
ISSN: 0894-069X

HISTORY: *Naval Research Logistics Quarterly*
                [ISSN: 0028-1441] V1-V33 (1954-1986)

EDITOR: Richard E. Rosenthal
                        Naval Research Logistics
                        Department of Operations Research
                        Code 55/Naval Postgraduate School
                        Monterey, CA 93943
MS REQUIREMENT: Four copies, dbl spaced, short abstract
(200 word maximum)
PUBLISHER: John Wiley
DATE FOUNDED: 1987 FREQUENCY: Bi-Monthly
PRICE: $195 MICROFORM: RPI
LANGUAGE: English
INDEXED/ABSTRACTED: Appl Mech Rev, Curr Cont, Eng Ind,
Math R, Risk Abstr, Sci Abstr
TARGET READER: Operations Researchers, Systems Analysts and
Programmers, Economists, Statisticians

Offers significant research papers that cover a broad range of operations research topics relevant to theoretical logistics. Articles are published on both theory and applications in key areas including mathematical statistics, economics, tactics, and strategy.

NEW ZEALAND MATHEMATICAL MAGAZINE
ISSN: 0549-0510

SEND MS TO: The Editors
                    New Zealand Mathematical Magazine

P. O. Box 26-226
Auckland 3. NEW ZEALAND
MS REQUIREMENT: Typewritten, dbl spaced
PUBLISHER: Auckland Mathematical Association, Inc.
DATE FOUNDED: 1963   FREQUENCY: 3/yr.
PRICE: $18.65   CIRC: 750
LANGUAGE: English   Back Issues Available
INDEXED/ABSTRACTED: Math R
TARGET READER: Mathematicians in all fields

Publishes articles of mathematical interest related to primary and secondary schools.

NIEUW ARCHIEF VOOR WISKUNDE
ISSN: 0028-9825

HISTORY: *Verslag van Het Verhandelde op de*
                Wetenschappelijke Vergaderingen (1842-1852)
                Archief V1-V3 (1856-1874)

MANAGING   M. Hazewinkel
EDITOR: Stichting Mathematisch Centrum
                Kruislaan 413
                P. O. Box 4079
                1009AB Amsterdam,
                THE NETHERLANDS
MS REQUIREMENT: Typewritten, dbl spaced, summary not to
exceed 150 words
PUBLISHER: Stichting Mathematisch Centrum
DATE FOUNDED: 1875   FREQUENCY: 3/yr.
PRICE: $59.06   CIRC: 2000
LANGUAGE: English, French, German, Dutch
INDEXED/ABSTRACTED: Math R, Ref Zh, Zent Math
TARGET READER: Mathematicians in all fields

Publishes mathematical papers that are of interest to a large proportion of the members of the Dutch Mathematical Society. Accepts introductory papers, survey papers, expository papers and historical notes for the general mathematical public.

NIHON TOKEI GAKKAISHI. *Journal of the Japan Statistical*
*Society**
ISSN: 0389-5602

PUBLISHER:	Nihon Teokei Gakkai		
DATE FOUNDED:	1970	FREQUENCY:	Semi-Annually
PRICE:	$40	LANGUAGE:	English, Japanese
CIRC:	1,500	Back Issues Available	

INDEXED/ABSTRACTED: Math R
TARGET READER: Statisticians

Publishes original papers in the areas of statistics.

NONLINEAR ANALYSIS
ISSN: 0362-546X

EDITOR:	V. Lakshmikantham		
	Department of Applied Math		
	Florida Institute of Tech		
	150 W. University Blvd.		
	Melbourne, FL 32901		
MS REQUIREMENT:	Duplicate, dbl spaced, list of key words,		
no abstract is required			
PUBLISHER:	Pergamon		
DATE FOUNDED:	1977	FREQUENCY:	24/yr.
PRICE:	$825	CIRC:	1000
LANGUAGE:	English	REPRINT:	UMI
MICROFORM:	UMI, MIM		

INDEXED/ABSTRACTED: Appl Meth Rev, Cam Sci Abstr, Compumath, Curr Cont, Math R, Sci Abstr
TARGET READER: Applied Mathematicians, Nonlinear Analysts

Publishes important research and expository papers and preliminary communications devoted to solving nonlinear problems in all areas of theory, methods, and applications of nonlinear analysis. Includes a software survey section.

NONLINEARITY
ISSN: 0951-7715

SEND MS TO:	Senior Managing Editor
	10P Publishing Ltd.
	Techno House, Redcliffe Way
	Bristol BS1 6NX, ENGLAND

MS REQUIREMENT: Triplicate, abstract, AMS Classification Numbers, accepts TEX form
PUBLISHER: American Institute of Physics
DATE FOUNDED: 1988   FREQUENCY: Quarterly
PRICE: $387   CIRC: 1200
LANGUAGE: English, French, German
INDEXED/ABSTRACTED: Comput Control Abstr, Elect Electron Abstr, Math R, Phys Abstr
TARGET READER: Mathematicians interested in nonlinear mathematics, physicists

Publishes papers on a wide range of nonlinear mathematics, mathematical and experimental physics, and other areas of the sciences where nonlinear phenomena are of fundamental importance.

NOTICES OF THE AMERICAN MATHEMATICAL SOCIETY
ISSN: 0002-9920

EDITOR:	Robert M. Fossum
	University of Illinois
	1409 West Green Street
	Urbana, IL 61801-2975
PUBLISHER:	American Mathematical Society

DATE FOUNDED:	1953	FREQUENCY:	10/yr.
PRICE:	$121	MICROFORM:	UMI
(Inst Memb)	$97	CIRC:	21000
(Indv Memb	$73	LANGUAGE:	English

INDEXED/ABSTRACTED: Comput Rev, Math R
TARGET READER: Mathematicians in all fields.

Reports meeting/conference announcements, current trends in scientific development, federal funding, news, and information of interest to mathematical community. Is one of the most widely read periodicals in the world dealing with matters of interest to the mathematical community.

NOTRE DAME JOURNAL OF FORMAL LOGIC
ISSN: 0029-4527

SEND MS TO:	Editors,
	Notre Dame Journal of Formal Logic
	P. O. Box 5
	Notre Dame, IN 46556

MS REQUIREMENT: Triplicate, dbl spaced, author's name on separate sheet, abstract at the beginning

PUBLISHER:		University of Notre Dame	
DATE FOUNDED:	1960	FREQUENCY:	Quarterly
PRICE:	(Inst) $45	CIRC:	825
	(Indv) $25	LANGUAGE:	English

Back Issues Available
INDEXED/ABSTRACTED: Math R, Phil Ind, Sci Abstr
TARGET READER: Logicians, Philosophers

Publishes articles in all areas of philosophical and mathematical logic, philosophy of language and formal semantics for natural languages, the philosophy, history and foundation of logic and mathematics, the philosophy and semantics for natural language.

*NUMERICAL ALGORITHMS**
ISSN: 1017-1398

EDITOR: C. Brezinski
                         Laboratoire d'Analyse Numérique et
                         d'Optimisation, UFR IEEA - M3
                         Universite des Sciences et
                         Techniques de Lile
                         Flandres-Artois 59655
                         Villeneuve d'Ascq Cedex, FRANCE
                         Tel: (33) 20 43 42 96
                         FAX: (33) 20 43 49 95
MS REQUIREMENT: Triplicate, dbl spaced, abstract, key words,
AMS (MOS) Classification Numbers, software programs must be
written in FORTRAN-77, Pascal or C-langugage
PUBLISHER: J. C. Baltzer AG
DATE FOUNDED: 1991   FREQUENCY: Quarterly
PRICE: $432.84 LANGUAGE: English
TARGET READER: Numerical Analysts, Computer Scientists

Publishes original and review papers on all aspects of numerical
algorithms: new algorithms, theoretical results, implementation,
numerical stability, complexity, parallel computing, subroutines and
parallel applications. Papers on computer algebra related to obtaining
of numerical results will be considered.

NUMERICAL FUNCTIONAL ANALYSIS AND OPTIMIZATION
ISSN: 0163-0563

EDITOR: M. Z. Nashed
                         University of Delaware
                         Newark, DE 19716
MS REQUIREMENT: Typewritten, 1½ line spacing, maximum
100-word abstract
PUBLISHER: Marcel Dekker
DATE FOUNDED: 1979   FREQUENCY: 10/yr.
PRICE: $265 MICROFORM: RPI
LANGUAGE: English, French, German, Russian
INDEXED/ABSTRACTED: Compumath, Curr Cont, Math R, Sci
Abstr

TARGET READER: Numerical Analysts

Publishes original research articles on the development and
application of functional analysis and operator-theoretic methods
within the field of numerical analysis, approximation theory,
optimization, control and system theory.

NUMERICAL METHODS FOR PARTIAL
DIFFERENTIAL EQUATIONS
ISSN: 0749-159X

EDITOR: George F. Pinder
                                 College of Engineering and Mathematics
                                 University of Vermont/Votey Bldg
                                 Burlington, VT 05405
MS REQUIREMENT: Original and 2 copies, dbl spaced, abstract
not to exceed 200 words
PUBLISHER: John Wiley
DATE FOUNDED: 1985 FREQUENCY: Quarterly
PRICE: $165 MICROFORM: RPI
LANGUAGE: English
INDEXED/ABSTRACTED: Appl Mech Rev, Math R
TARGET READER: Applied Mathematicians, Engineers, Physicists

Publishes papers of new developments in the use of numerical methods
for the solution of partial differential equations in the areas of science
and engineering.   The main focus is on the exposition of novel
methodology.

NUMERISCHE MATHEMATIK
ISSN: 0029-599X

EDITOR: R. S. Varga
                                 Institute for Computational
                                   Mathematics
                                 Kent State University
                                 Kent, OH 44242

MS REQUIREMENT: Triplicate, short summary, 1991 AMS
Subject Classification Numbers
PUBLISHER: Springer-Verlag
DATE FOUNDED: 1959 FREQUENCY: 16/yr.
PRICE: $1004 MICROFORM: UMI
REPRINT: ISI   LANGUAGE: English, French,
                                              German
INDEXED/ABSTRACTED: Appl Mech Rev, Curr Cont, Math R,
Zent Math
TARGET READER: Numerical Analysts

Publishes papers presenting significantly new and important
developments in all areas of numerical analysis.   Provides for
dissemination of original contributions dealing with mathematical topics
which arise in contemporary numerical computation including
optimization of parallel computers.

OPERATIONS RESEARCH
ISSN: 0030-364X

HISTORY: *Journal of the Operations Research*
                 Society of America
                 [ISSN: 0096-3984] V1-V3 (1952-1955)

EDITOR: H. Donald Ratliff
                          School of Industrial and Systems
                            Engineering
                          Georgia Tech
                          Atlanta, GA 30332
MS REQUIREMENT: Four copies, dbl spaced, abstract not to
exceed 200 words, include up to three appropriate subject
classifications
PUBLISHER: Operations Research Society of America
DATE FOUNDED: 1956 FREQUENCY: Bi-Monthly
PRICE: (Inst) $110 CIRC: 10,100
              (Indv) $74   LANGUAGE: English
MICROFORM: KTO BACK ISSUES: Kraus Reprint
INDEXED/ABSTRACTED: Appl Mech Rev, Bus Ind, Chem Abstr,

Eng Ind, Math R, SSCI
TARGET READER: Mathematicians interested in operations
research, System Scientists

Publishes papers on operations research and management science work
of interest to the operations research practioner and researcher in three
substantive categories: methods, data based operational science, and the
practices of OP. Includes papers reporting underlying data based
principles of operational science, observations and modeling of
operating systems, contributions to the methods and models of OP, case
histories of applications, etc.

OPTIMIZATION
ISSN: 0233-1934

HISTORY: *Mathematische Operationsforschung und Statistik*
                [ISSN: 0047-6277] Bd1-Bd7 (1970-1976)
                Optimization [ISSN: 0323-3898] V8-V15
                  (1977-1984)

EDITOR: K. H. Elster, TH Ilmenau
                            Institut für Mathematik
                            Am Ehrenberg
                            0-6300 Ilmenau, GERMANY
MS REQUIREMENT: Original and two copies, dbl spaced, 150
word summary, AMS 1980 Subject Classification Numbers, key
words
PUBLISHER: Akademie-Verlag
DATE FOUNDED: 1985   FREQUENCY: Quarterly
PRICE: $182.33 LANGUAGE: English, French,
                                                German, Russian
INDEXED/ABSTRACTED: Comput Control Abstr, Elect Electron
Abstr, Math R, Phys Abstr
TARGET READER: Operations Researchers, Computer Scientists

Publishes papers on mathematical programming and operations research.

OSAKA JOURNAL OF MATHEMATICS
ISSN: 0030-6126

HISTORY: *Journal of the Institute of Polytechnics,*
                Osaka City University.
                Series A. Mathematics
                V1-V11 (1950-1960)
                Journal of Mathematics,
                Osaka City University
                [ISSN: 0449-2773] V12-V14 (1961-1963)
        merged with
                Osaka Mathematical Journal,
                V1-V15 (1949-1963)

PUBLISHER: Osaka University
DISTRIBUTED BY: Kinokyniya Co., Ltd.
DATE FOUNDED: 1964 FREQUENCY: Quarterly
PRICE: $257 LANGUAGE: English, French
INDEXED/ABSTRACTED: Compumath, Math R, Sci Cit Ind
TARGET READER: Mathematicians in all fields

Publishes research papers on all aspects of mathematics.

PACIFIC JOURNAL OF MATHEMATICS
ISSN: 0030-8730

MANAGING   V. S. Varadarajan
EDITOR: University of California
                        Los Angeles, CA 90024-1555-05
MS REQUIREMENT: Triplicate, dbl spaced, 1991 Mathematics
Subject Classification Numbers, brief synopsis
PUBLISHER: Pacific Journal of Mathematics
DATE FOUNDED: 1951 FREQUENCY: Monthly
PRICE: (Inst) $190 (except July and August)
                (Indv) $95   LANGUAGE: English
CIRC: 1600 Back Issues Available

INDEXED/ABSTRACTED: Compumath, Curr Cont, Math R, Zent
Math
TARGET READER: Mathematicians in all fields

Publishes articles in pure and applied mathematics.

PANAMERICAN MATHEMATICAL JOURNAL

EDITOR: Ram U. Verma
                           Panamerican Mathematical Journal
                           Department of Mathematics
                           University of Central Florida
                           Orlando, FL 32816
MS REQUIREMENT: Duplicate
PUBLISHER: Panamerican Mathematical Journal
                           Department of Mathematics
                           University of Central Florida
DATE FOUNDED: 1991   FREQUENCY: Quarterly
PRICE: (Inst) $100   LANGUAGE: English
                (Indv) $50
TARGET READER: Mathematicians in all fields

Publishes high quality articles in all areas of mathematical science. A
substantial introductory section giving the motiviation, background, and
significance of the results is needed. The introductory section should
be addressed to a broad spectrum of mathematical enthusiasts.

PHILOSOPHIA METHEMATICA
ISSN: 0031-8019

EDITOR: J. Fang, Prof. Emer.
                           Old Dominion University
                           Norfolk, VA 23529
MS REQUIREMENT: Original typewritten and one copy, dbl
spaced, wide margin
PUBLISHER: J. Fang

DATE FOUNDED:		1964	FREQUENCY:	Semi-Annually
2nd Series:		1986	CIRC:	600
PRICE:	(Inst)	$35	LANGUAGE:	English, French
	(Indv)	$20		German

INDEXED/ABSTRACTED: Math R, Phil Ind, Zent Math
TARGET READER: Mathematicians, Philosophers

Publishes papers of inter-, cross-, multi-, or even trans - disciplinary studies in the current explosion of knowledge in philosophy, history and sociology of mathematics.

PI MU EPSILON JOURNAL
ISSN: 0031-952X

SEND MS TO:		Editor, Pi Mu Epsilon Journal		
		Mathematics & Computer		
		Science Department		
		Macalester College		
		St. Paul, MN 55105		
MS REQUIREMENT:		Duplicate, must be correct and honest		
PUBLISHER:		Pi Mu Epsilon		
DATE FOUNDED:		1949	FREQUENCY:	2/yr.
(Subscription runs			MICROFORM:	UMI
for 5 yrs/1 vol)			REPRINT	UMI
PRICE:	(Inst)	$30	CIRC:	5000
	(Memb)	$20	LANGUAGE:	English

INDEXED/ABSTRACTED: Math R
TARGET READER: Undergraduate Mathematics Students, Mathematics Teachers

Dedicated to undergraduate and beginning college students interested in mathematics. Student publications are given priority. Each year, at least five students are given awards for their papers.

PROBABILITY THEORY AND RELATED FIELDS
ISSN: 0178-8051

HISTORY: *Zeitschrift für Wahrscheinlichkeitstheorie und*
             Verwandte Gebiete [ISSN: 0044-3719] V1-V70
             (1962-1985)

MANAGING
EDITOR: H. Rost
MS REQUIREMENT: Duplicate
PUBLISHER: Springer-Verlag
DATE FOUNDED: 1986   FREQUENCY: Irr
PRICE: $1598.99 MICROFORM: UMI
LANGUAGE: English (Mainly), French, German
INDEXED/ABSTRACTED: Curr Cont, Curr Math Publ, Math R,
Zent Math
TARGET READER: Probability Theorists, Probabilists

Publishes research papers in advanced probability theory and other
fields such as optimization theory, statistical mechanics, or analytic
number theory when they are closely connected with basic problems in
probability theory.

PROCEEDINGS OF THE AMERICAN MATHEMATICAL SOCIETY
ISSN: 0002-9939

EDITOR: Editorial Department
                       American Mathematical Society
                       P. O. Box 6248
                       Providence, RI 02940-6248
MS REQUIREMENT: Duplicate, dbl spaced, short notes - not to
exceed two printed pages, accept papers of electronic form
PUBLISHER: American Mathematical Society
DATE FOUNDED: 1950 FREQUENCY: Monthly
PRICE: $470 LANGUAGE: English
       (Inst Memb) $376 CIRC: 1900
       (Indv Memb) $282 MICROFORM: UMI
Back Volumes Available:AMS
INDEXED/ABSTRACTED: Compumath, Curr Cont, Math R,

TARGET READER: Pure and Applied Mathematicians

Devoted entirely to research in pure and applied mathematics, principally to the publication of original papers of moderate length. Tends to publish short papers usually less than 10 typewritten pages. A department called Shorter Notes was established for the purpose of publishing very short papers of an unusually elegant and polished character for which there is normally no other outlet.

PROCEEDINGS OF THE EDINBURGH MATHEMATICAL SOCIETY
ISSN: 0013-0915

HISTORY   Series I comprised 44 volumes,
                 published from 1884-1926.
                 Series II commenced in 1927.

SEND MS TO: Secretary (or Editor)
                         Edinburgh Mathematical Society
                         James Clerk Maxwell Bldg
                         The King's Buildings
                         Edinburgh, EH9 3JZ SCOTLAND
MS REQUIREMENT: Duplicate, dbl spaced, AMS 1980
Mathematics Subject Classification Numbers
PUBLISHER: Scottish Academic Press
DATE FOUNDED: 1883   FREQUENCY: 3/yr.
PRICE: $150 LANGUAGE: English
Back Issues Available from Kraus Reprint Ltd.
INDEXED/ABSTRACTED: Compumath, Curr Cont, Math R
TARGET READER: Mathematicians in all fields

Covers a wide range of topics in both pure and applied mathematics.

PROCEEDINGS OF THE INDIAN ACADEMY OF SCIENCES
ISSN: 0253-4142

HISTORY: *Proceedings of the Indian*
                  Academy of Sciences. Section A
                  [ISSN 0370-0089]
                  V1-V85 (1934-June 1977)
                  Proceedings. A,
                  V86-87 (July, 1977-1978)
                  Proceedings. A, Mathematical Sciences.
                  V88 (1979)

EDITOR: S. G. Dani
                            Tata Institute of Fundamental Research
                            Bombay, INDIA
PUBLISHER: Indian Academy of Sciences
DATE FOUNDED: 1980 FREQUENCY: 3/yr.
PRICE: $75   CIRC: 1000
MIRCOFORM: UMI   REPRINT: UMI
LANGUAGE: English
INDEXED/ABSTRACTED: Compumath, Curr Cont
TARGET READER: Mathematicians in all fields

Publishes papers on pure and applied mathematics.

PROCEEDINGS OF THE LONDON MATHEMATICAL SOCIETY
ISSN: 0024-6115

HISTORY: *1st Series* (1865-1903)
                  2nd Series (1903-1953)

EDITOR: Various Editors on Different Subjects
MS REQUIREMENT: Duplicate, dbl spaced, AMS Subject
Classification Numbers
PUBLISHER: Oxford University Press
DATE FOUNDED: 1954 FREQUENCY: 6/yr.
PRICE: $480 CIRC: 1400
MICROFORM: UMI   LANGUAGE: English, French,
                                             German
INDEXED/ABSTRACTED: Appl Mech Rev, Compumath, Curr

Cont, Math R, Sci Cit Ind
TARGET READER: Mathematicians interested in analysis and
differential equations.

Publishes longer research papers for the London Mathematical Society
in the fields of real and complex analysis, differential equations and
related areas.

PROCEEDINGS OF THE NATIONAL ACADEMY OF SCIENCES
ISSN: 0027-8424

EDITOR: Lawrence Bogorad
MS REQUIREMENT: Only members or foreign associates of the
Academy may submit papers, a member may submit up to six
manuscripts in each calendar year. Research papers may not exceed
5 printed pages. Duplicate of the manuscript and one extra copy of
both the title page and abstract, five key terms.
PUBLISHER: National Academy of Sciences
DATE FOUNDED: 1915 FREQUENCY: Semi-Monthly
PRICE: (Inst) $380 CIRC: 10,000
                (Indv) $210 MICROFORM: UMI
                (Stud) $80   LANGUAGE: English
PUBLICATION CHARGES: $50/page
INDEXED/ABSTRACTED: Biol Abstr, Chem Abstr, Excerpta Med,
Ind Med, Math R, Sci Abstr
TARGET READER: Scientists in various fields

Publishes articles describing original work in the sciences that are of
interest to a wide audience.

PROCEEDINGS OF THE ROYAL SOCIETY OF EDINBURGH.
SECTION A: MATHEMATICS
ISSN: 0308-2105

HISTORY: *Proceedings of the Royal Society of Edinburgh*

[ISSN: 0080-4541] V1-V60 (1832-1940)
Proceedings of the Royal Society of Edinburgh,
Section A: Mathematical and Physical Sciences
[ISSN: 0080-4541] V61-V67 (1941-1967)
Proceedings. Section A, Mathematical and
Physical Science [ISSN: 0080-4541] V68-V71
(1968-1974)

SEND MS TO: Publications Manager
                     The Royal Society of Edinburgh
                     22 George Street
                     Edinburgh, EH2 2PQ, SCOTLAND
MS REQUIREMENT: Triplicate, dbl spaced, synopsis not
to exceed 200 words
PUBLISHER: Royal Society of Edinburgh
DATE FOUNDED: 1974 FREQUENCY: Bi-Monthly
PRICE: $308 CIRC: 1000
LANGUAGE: English
INDEXED/ABSTRACTED: Appl Mech Rev, Biol Abstr, Chem
Abstr, Compumath, Curr Cont, Eng Ind, Math R, Sci Abstr
TARGET READER: Mathematicians in all fields

Publishes papers in all areas of mathematics.

PROCEEDINGS OF THE STEKLOV INSTITUTE OF MATHEMATICS
ISSN: 0081-5438

EDITOR: Ben Silver
PUBLISHER: American Mathematical Society
DATE FOUNDED: 1967 FREQUENCY: Quarterly
PRICE: (Inst) $480 CIRC: 550
              (Inst Memb) $384   LANGUAGE:English
INDEXED/ABSTRACTED: Math R
TARGET READER: Mathematicians in all fields

This is a cover-to-cover English translation of Trudy Ordena Lenina
Matematicheskogo Institua imeni V.A. Steklova of the Academy of

Sciences of the U.S.S.R. Each issue contains either one book-length article or a collection of articles pertaining to the same topic.

QUARTERLY JOURNAL OF MATHEMATICS. Oxford
Series
ISSN: 0033-5606

HISTORY:	*Cambridge Mathematical Journal,*
	V1-V4 (1837-1845)
	Cambridge and Dublin Mathematical Journal,
	V1-V9 (1846-1854)
	Quarterly Journal of Pure and Applied
	Mathematics, V1-V50 (1855-1927)

(merged with)

Messenger of Mathematics, V1-V58 (1871-1929)

EDITORS:	Mathematical Institute
	24-29 St. Giles
	Oxford, ENGLAND
MS REQUIREMENT:	Duplicate, generous spacing
PUBLISHER:	Oxford University Press
DATE FOUNDED:	1930 FREQUENCY: Quarterly
PRICE:	$125 CIRC: 1200
MICROFORM:	UMI LANGUAGE: English

INDEXED/ABSTRACTED: Compumath, Curr Cont, Math R, Sci Cit Ind, Stat Theory Meth Abstr
TARGET READER: Mathematicians in all fields

Publishes articles of pure mathematics and its applications as well as the main branches of algebra, analysis, combinatorics, and topology.

QUARTERLY JOURNAL OF MECHANICS AND APPLIED MATHEMATICS
ISSN: 0033-5614

EDITOR: L. M. Hocking

Department of Math
University College, London
Gower Street
London, WC1E 6BT, ENGLAND

MS REQUIREMENT: Typewritten, dbl spaced, summary not to
exceed 300 words
PUBLISHER: Oxford University Press
DATE FOUNDED: 1948   FREQUENCY: 4/yr.
PRICE: $165   CIRC: 1400
MICROFORM: UMI   LANGUAGE: English
INDEXED/ABSTRACTED: Appl Mech Rev, Chem Abstr, Curr
Cont, Eng Ind, Math R, Sci Cit Ind
TARGET READER: Physicists, Engineers, Mathematicians

Publishes original articles in the general field of mechanics, particularly
theoretical mechanics, classical electromagnetism, nonlinear dynamics
and combined fields such as magnetohydro-numerical methods.

QUARTERLY OF APPLIED MATHEMATICS
ISSN: 0033-569X

EDITORIAL   Brown University
OFFICE: Box F
                       Providence, RI 02912
MS REQUIREMENT: Duplicate, dbl spaced
PUBLISHER: Brown University
                       American Mathematical Society
DATE FOUNDED: 1943   FREQUENCY: Quarterly
PRICE: $55   CIRC: 1600
REPRINT: UMI   LANGUAGE: ENGLISH
Back Issues Avalaible   MICROFORM: UMI
INDEXED/ABSTRACTED: Appl Mech Rev, Biol Abstr, Chem
Abstr, Compumath, Curr Cont, Eng Ind, Math R
TARGET READER: Applied Mathematicians, Research Scientists,
Engineers

Publishes original papers in applied mathematics which have an intimate
connection with applications. Welcomes contributions which will be of
interest both to mathematicians as well as scientists and/or engineers.

RANDOM STRUCTURES AND ALGORITHMS
ISSN: 1042-9832

EDITOR: Joel Spencer
                           Courant Instititute of Mathematical
                           Sciences
                           New York University
                           251 Merces Street
                           New York, NY 10012
MS REQUIREMENT: Original and one copy, dbl spaced,
abstract not to exceed 50 words, 5 key words
PUBLISHER: John Wiley
DATE FOUNDED: 1990 FREQUENCY: Quarterly
PRICE: $169 LANGUAGE: English
Back Issues Available
TARGET READER: Mathematicians, Computer Scientists,
Operations Researchers

Covers the latest research on discrete random structures. Presents
applications of such research to problems in combinatorics and
computer science. Provides a useful forum for ideas on future studies
in randomness.

REAL ANALYSIS EXCHANGE
ISSN: 0147-1937

SEND MS TO: Real Analysis Exchange
                           Department of Math
                           Michigan State University
                           East Lansing, MI 48824-1027
MS REQUIREMENT: Accepts camera-ready form and any form
of TEX, for detail information contact Prof. Paul Humk, Department
of Math, St. Olaf College, Northfield, MN 55057

PUBLISHER: Michigan State University
DATE FOUNDED: 1976 FREQUENCY: Semi-Annually
PRICE: (Inst) $33 CIRC: 425
             (Indv) $20 LANGUAGE: English
Back Issues Available
INDEXED/ABSTRACTED: Math R, Ref Zh, Zent Math
TARGET READER: Analysts, Set Theorists

Publishes papers concerning areas of Real Analysis in four sections: Topical Surveys, Research Articles, Inroads, and Queries. Topical Surveys consist of articles concerning one area of current research activity. Research Articles include research in real analysis, functions of one or more real variables and real set theory. Inroads present new and simple proofs of well-known theorems, or simple and interesting consequences of known results. Queries include problems with appropriate bibliographical and historical information.

REVIEW OF ECONOMICS AND STATISTICS
ISSN: 0034-6535

EDITOR: Hendrik S. Houthakker
SEND MS TO: Review of Economics and Statistics
                  M-8 Littauer Center,
                  Cambridge, MA 02138
MS REQUIREMENT: Triplicate
PUBLISHER: Elsevier
DATE FOUNDED: 1966 FREQUENCY: 4/yr.
PRICE: (Inst) $135 MICROFORM: UMI, RPI
             (Indv) $57.50 CIRC: 5400
LANGUAGE: English
INDEXED/ABSTRACTED: CREJ, Curr Cont, Manage Cont, SSCI, Soc Sci Ind
TARGET READER: Economists, Statisticians, Bankers

Publishes theoretical, empirical and statistical analysis that characterizes modern economics.

ROCKY MOUNTAIN JOURNAL OF MATHEMATICS
ISSN: 0035-7596

MANAGING   John N. McDonald
EDITOR: Department of Math
               Arizona State University
               Tempe, AZ 85287-1904
MS REQUIREMENT:Duplicate, dbl spaced, accept TEX form with
Duplicate of manuscript
PUBLISHER: Rocky Mountain Mathematics
             Consortium
DATE FOUNDED   1971 FREQUENCY: Quarterly
PRICE: (Inst) $350 CIRC: 500
             (Indv) $175 Back Issues Available
LANGUAGE: EnglishPUBL CHARGES: $35
                                  (to the institution or grant)
INDEXED/ABSTRACTED: Compumath, Math R, Sci Abstr
TARGET READER: Mathematicians in all fields

Publishes both research and expository articles in mathematics and
solicits well-written survey articles.

RUSSIAN MATHEMATICAL SURVEYS
ISSN: 0036-0279

English translation of: *Uspekhi Matematicheskikh Nauk*
       [ISSN: 0042-1316]

EDITOR: E. J. F. Primrose
             12 King Road
             Leicester, LE2 3RR, ENGLAND
PUBLISHER: London Mathematical Society
DATE FOUNDED: 1960   FREQUENCY: Bi-Monthly
PRICE: $560.29   MICROFORM: MIM
LANGUAGE: English
INDEXED/ABSTRACTED: Compumath, Math R
TARGET READER: Mathematicians in all fields

A translation of the survey articles, biographical material, and communications of the Moscow Mathematical Society in Uspekhi Matematicheskikh Nauk.

*SCANDINAVIAN JOURNAL OF STATISTICS**
ISSN: 0303-6898

PUBLISHER: Allen Press
DATE FOUNDED: 1974 FREQUENCY: Quarterly
PRICE: $119 LANGUAGE: English
Back Issues Available
INDEXED/ABSTRACTED: Comput Control, Elect Electron Abstr, Math R
TARGET READER: Probabilists, Statisticians

Publishes research papers in theoretical and applied statistics. Also covers well motivated papers on relevant aspects of probability and other fields.

SEMIGROUP FORUM
ISSN: 0037-1912

EDITOR: J. A. Goldstein
     Department of Math
     Tulane University
     New Orleans, LA 70118
MS REQUIREMENT:Encouraged to submit in TEX form, for typewritten, 1½ line spaced, dbl space for mathematical symbols
PUBLISHER: Springer-Verlag
DATE FOUNDED: 1970 FREQUENCY: 6/yr.
PRICE: (Inst) $203 MICROFORM: UMI
     (Indv) $120 REPRINT: ISI
LANGUAGE: English (preferred), French, German, Russian
INDEXED/ABSTRACTED: Compumath, Curr Cont, Math R, Zent Math
TARGET READER: Semigroup Theorists

Serves as a platform for information transmissions on current research in semigroup theory. The scopes of the journal are: algebraic semigroups, topological semigroups, partially ordered semigroups, transformation semigroups, semigroups or operators, and applications of semigroup theory to other disciplines such as ring theory, category theory, automata, logic, etc. It covers original research papers, survey articles, research announcements, short notes research problems and announcements of conferences, seminars, and symposia on semigroup theory.

SIAM JOURNAL ON APPLIED MATHEMATICS
ISSN: 0036-1399

HISTORY: *Journal of the Society for Industrial And Applied
         Mathematics* [ISSN: 0368-4245] V1-V13 (1953-1965)

MANAGING
EDITOR: James P. Keener
SEND MS TO: Editor
                          SIAM Publications
                          Box 7541
                          Philadelphia, PA 19101
MS REQUIREMENT: Five copies, dbl spaced, abstracts not to
exceed 250 words, list of key words, AMS (MOS) Subject
Classification, TEX, LaTEX, and AMS-TEX are accepted
PUBLISHER: SIAM
DATE FOUNDED: 1966   FREQUENCY: Bi-Monthly
PRICE: (Inst) $195 CIRC: 3000
              (Indv) $48   LANGUAGE: English
Back Issues Available
INDEXED/ABSTRACTED: Appl Mech Rev, Chem Abstr,
Compumath, Comput Rev, Cyb Abstr, Eng Ind, Math R, Sci Abstr
TARGET READER: Applied Mathematicians, Engineers, Medical
Scientists, Biologists, Natural Scientists

Publishes research articles in mathematical methods and their applications in the physical, engineering, biological, and medical sciences.

SIAM JOURNAL ON COMPUTING
ISSN: 0097-5397

MANAGING
EDITOR: Andrew C. Yao
SEND MS TO: Editor
                           SIAM Publications
                           Box 7541
                           Philadelphia, PA   19101
MS REQUIREMENT: Five copies, dbl spaced, abstracts not to
exceed 250 words, list of key words, AMS (MOS) Subject
Classification, TEX, LaTEX, and AMS-TEX are accepted
PUBLISHER: SIAM
DATE FOUNDED: 1972 FRENQUENCY   Bi-Monthly
PRICE: (Inst) $185 CIRC: 1982
                   (Indv) $48   LANGUAGE: English
Back Issues Available
INDEXED/ABSTRACTED: Appl Mech Rev, Compumath, Comput
Rev, Cyb Abstr, Math R, Sci Abstr
TARGET READER: Mathematicians, Scientific and Computer
Professionals

Contains research articles in the application of mathematics to the
problems of computer science and the nonnumerical aspects of
computing.

SIAM JOURNAL ON CONTROL AND OPTIMIZATION
ISSN: 0363-0129

HISTORY: *Journal of the Society for Industrial and Applied*
                Mathematics. Series A: On Control
                [ISSN: 0887-4603] V1-V3 (1962-1965)
                SIAM *Journal on Control*
                [ISSN: 0036-1402] V4-V13 (1966-1975)

MANAGING
EDITOR: Jan C. Williams

SEND MS TO: Editor
                         SIAM Publications
                         Box 7541
                         Philadelphia, PA 19101
MS REQUIREMENT: Five copies, dbl spaced, abstracts not to
exceed 250 words, list of key words, AMS (MOS) Subject
Classification, TEX, LaTEX, and AMS-TEX are accepted
PUBLISHER: SIAM
DATE FOUNDED: 1976   FREQUENCY: Bi-Monthly
PRICE: (Inst) $225   CIRC: 2000
                 (Indv) $48   LANGUAGE: English
Back Issues Available
INDEXED/ABSTRACTED: Appl Mech Rev, Compumath, Cyb
Abstr, Math R, Sci Abstr
TARGET READER: Engineers, System Scientists, Mathematicians
interested in control and optimization

Publishes research articles in the mathematical theory of control and its
applications and the associated areas of systems theory and
optimization.

SIAM JOURNAL ON DISCRETE MATHEMATICS
ISSN: 0895-4801

HISTORY: SIAM *Journal on Algebraic and Discrete Methods*
                [ISSN: 0196-5212] V1-V8 (1980-1987)

MANAGING
EDITOR: William T. Trotter
SEND MS TO: Editor
                         SIAM Publications
                         Box 7541
                         Philadelphia, PA 19101
MS REQUIREMENT: Five copies, dbl spaced, abstracts not to
exceed 250 words, list of key words, AMS (MOS) Subject
Classification, TEX, LaTEX, and AMS-TEX are accepted
PUBLISHER: SIAM
DATE FOUNDED: 1988   FREQUENCY: Quarterly
PRICE: (Inst) $175   CIRC: 800

                    (Indv) $48   LANGUAGE: English
Back Issues Available
INDEXED/ABSTRACTED: Compumath, Int Abstr Oper Res, Math
R
TARGET READER: Mathematicians in various fields, such as
coding theory, and discrete mathematics, Computer Scientists

Contains research articles on a broad range of topics from pure and
applied mathematics including combinatorics and graph theory, discrete
optimization and operations research, theoretical computer science,
coding and communication theory, game theory and mathematical
modeling.

SIAM JOURNAL ON MATHEMATICAL ANALYSIS
ISSN: 0036-1410

MANAGING
EDITOR: Jerry L. Bona
SEND MS TO: Editor
                        SIAM Publications
                        Box 7541
                        Philadelphia, PA 19101
MS REQUIREMENT: Five copies, dbl spaced, abstracts not to
exceed 250 words, list of key words, AMS (MOS) Subject
Classification, TEX, LaTEX, and AMS-TEX are accepted
PUBLISHER: SIAM
DATE FOUNDED: 1970   FREQUENCY: Bi-Monthly
PRICE: (Inst) $285   CIRC: 1372
                (Indv) $48   LANGUAGE: English
Back Issues Available
INDEXED/ABSTRACTED: Compumath, Comput Rev, Math R, Sci
Abstr
TARGET READER: Analysts, Engineers

Focuses on parts of classical and modern articles that have direct or
potential application to the natural sciences and engineering. Papers fall
into two broad categories: analysis of interesting problems associated

with realistic mathematical models for natural phenomena, and contributions in a substantial way to the general analytical information and techniques.

SIAM JOURNAL ON MATRIX ANALYSIS AND APPLICATIONS
ISSN: 0895-4798

HISTORY: *SIAM Journal on Algebraic and Discrete Methods*
                [ISSN: 0196-5212] V1-V8 (1980-1987)

MANAGING   Gene H. Golub
EDITOR: SIAM Publications
                    P. O. Box 7541
                    Philadelphia, PA 19101
MS REQUIREMENT: Five copies, dbl spaced, abstract not to exceed 250 words, key words, AMS(MOS) Subject Classifications, camera ready copy and TEX files encouraged
PUBLISHER: SIAM
DATE FOUNDED: 1988 FREQUENCY: Quarterly
PRICE: $130 LANGUAGE: English
INDEXED/ABSTRACTED: Comput Control Abstr, Elect Electron Abstra, Phys Abstr
TARGET READER: Analysts, Applied Mathematicians interested in Matrix Theory

Contains research articles on the application of matrix analysis to areas such as Markov chains, networks, signal processing, systems and control theory, mathematical programming economic and biological modeling, etc.

SIAM JOURNAL ON NUMERICAL ANALYSIS
ISSN: 0036-1429

HISTORY: *Journal of the Society for Industrial and Applied Mathematics. Series B: Numerical Analysis*
                [ISSN: 0887-459X] V1-V2 (1964-1965)

MANAGING
EDITOR: Mitchell Luskin
SEND MS TO: Editor
                           SIAM Publications
                           Box 7541
                           Philadelphia, PA   19101
MS REQUIREMENT: Five copies, dbl spaced, abstracts not to
exceed 250 words, list of key words, AMS (MOS) Subject
Classification, TEX, LaTEX, and AMS-TEX are accepted
PUBLISHER: SIAM
DATE FOUNDED: 1966   FREQUENCY: Bi-Monthly
PRICE: $205   CIRC: 3000
LANGUAGE: English Back Issues Available
INDEXED/ABSTRACTED: Appl Mech Rev, Compumath, Comput
Rev, Math R, Sci Abstr
TARGET READER: Numerical Analysts

Contains research articles on the development and analysis of numerical
methods including their convergence, stability, and error analysis as well
as related results in functional analysis and approximation theory.
Computational experiments and new types of numerical applications are
included.

SIAM JOURNAL ON OPTIMIZATION
ISSN: 1052-6234

MANAGING
EDITOR: J. E. Dennis
PUBLISHER: SIAM
DATE FOUNDED: 1991   FREQUENCY: Quarterly
PRICE: (Inst) $150 LANGUAGE: English
              (Memb) $40
TARGET READER: Applied Mathematicians, Computer Scientists,
Engineers

Contains research and expository articles on the theory and practice of
optimization, and papers that link optimization theory with
computational practices and applications.

SIAM JOURNAL ON SCIENTIFIC AND STATISTICAL COMPUTING
ISSN: 0196-5204

MANAGING
EDITOR: Linda Petzold
SEND MS TO: Editor
                       SIAM Publications
Box 7541   Philadelphia, PA 19101
MS REQUIREMENT: Five copies, dbl spaced, abstracts not to exceed 250 words, list of key words, AMS (MOS) Subject Classification, TEX, LaTEX, and AMS-TEX are accepted
PUBLISHER: SIAM
DATE FOUNDED: 1980 FREQUENCY: Bi-Monthly
PRICE: (Inst) $195 CIRC: 1500
                (Indv) $48   LANGUAGE: English
Back Issues Available
INDEXED/ABSTRACTED: Compumath, Cyb Abstr, J Cont Quant Meth, Math R, Sci Abstr
TARGET READER: Applied Mathematicians, Computer Scientists

Publishes research papers on those techniques of scientific computation concerned with the solution of continuous or statistical models. Welcomes papers on new algorithms for current architectures or developing new architectures.

*SIAM NEWS**

HISTORY: *Formerly SIAM Newsletter*
               [ISSN: 0036-1437] V1-V5 (1968-1972)

PUBLISHER: SIAM
DATE FOUNDED: 1973 FREQUENCY: Bi-Monthly
PRICE: $15   CIRC: 8500
LANGUAGE: English
INDEXED/ABSTRACTED: Math R

TARGET READER: Mathematicians, Computer Scientists, Scientists interested in mathematics

This newsjournal forges links between pure and applied mathematicians, mathematics teachers, computer scientists, engineers, statisticians, physicists, professionals involved in scientific computing, and other users of mathematics working in virtually every area of science.

SIAM REVIEW
ISSN: 0036-1445

MANAGING EDITOR:	Paul W. Davis
SEND MS TO:	Editor
	SIAM Publications
	Box 7541
	Philadelphia, PA 19101

MS REQUIREMENT: Five copies, dbl spaced, abstracts not to exceed 250 words, list of key words, AMS (MOS) Subject Classification, TEX, LaTEX, and AMS-TEX are accepted

PUBLISHER:	SIAM		
DATE FOUNDED:	1959	FREQUENCY:	Quarterly
PRICE:	$120	CIRC:	8500
Back Issues Available		LANGUAGE:	English

INDEXED/ABSTRACTED: Appl Mech Rev, Compumath, Comput Rev, Elect Electron Abstr, Excerpta Med, Math R
TARGET READER: Applied Mathematicians, Statisticians

Contains primary expository and survey papers as well as occasional essays on topics of interest to applied mathematicians. This journal also publishes reviews.

SIBERIAN MATHEMATICAL JOURNAL
ISSN: 0037-4466

PUBLISHER: Consultants Bureau

DATE FOUNDED: 1966 FREQUENCY: Bi-Monthly
PRICE: $1125 LANGUAGE: English
INDEXED/ABSTRACTED: Comput Inf Syst Abstr, Curr Cont,
Math R, Zent Math
TARGET READER: Mathematicians in all fields

Translation of Sibirskii Matematicheskii Zhurnal, a publication of the
Siberian Branch of the Academy of Science of USSR, Novosibirsk.

SOVIET JOURNAL OF CONTEMPORARY MATHEMATICAL ANALYSIS
ISSN: 0735-2719

English translation of *Akademiya Nauk Armyanskoi S.S.R.*
Izvestiya. Seriya Mathematika [ISSN: 0002-3043]

EDITOR: M. M. Dzhrbashyan
PUBLISHER: Allerton Press
DATE FOUNDED: 1979 FREQUENCY: Bi-Monthly
PRICE: $625 LANGUAGE: English
INDEXED/ABSTRACTED: Math R
TARGET READER: Analysts

Is a cover-to-cover translation from Russian papers in the field of
mathematical analysis published by the Armenian Academy of Sciences.

SOVIET JOURNAL OF NUMERICAL ANALYSIS AND MATHEMATICAL MODELLING
ISSN: 0169-2895

EDITOR: G. I. Marchuk
PUBLISHER: VSP
DATE FOUNDED: 1986 FREQUENCY: Bi-Monthly
PRICE: $631 LANGUAGE: English
INDEXED/ABSTRACTED: Math R

TARGET READER: Numerical Analysts, Physicists

Provides English translations of selected new, original Soviet papers on the theoretical aspects of numerical analysis and the application of mathematical methods to simulation and modelling. Papers are selected on the basis of their high scientific standard, innovative approach, and topical interest.

SOVIET MATHEMATICS - IZ. VUZ
ISSN: 0197-7156

Cover to cover translation from *Russian Soviet Mathematica*
        [ISSN: 0021-3446].

EDITOR: A. V. Sul'din
PUBLISHER: Allerton Press
DATE FOUNDED: 1974   FREQUENCY: Monthly
PRICE: $750   LANGUAGE: English
INDEXED/ABSTRACTED: Comput Control Abstr, Elect Electron Abstr, Math R, Phys Abstr, Stat Theory Meth Abstr
TARGET READER: Mathematicians in all fields

Publishes papers in the area of pure and applied mathematics.

SOVIET MATHEMATICS-DOKLADY
ISSN: 0197-6788

HISTORY: *Soviet Mathematics*
                [ISSN: 0038-5573] V1-V19 (1960-1978)
                Translation of the Mathematics Section of
                *The Doklady Akademii Nauk SSSR, The Reports
                of the Academy of Sciences of the U.S.S.R.*

EDITOR: Lev J. Leifman
PUBLISHER: American Mathematical Society
DATE FOUNDED: 1979   FREQUENCY: 6/yr.
PRICE: (Inst) $602   CIRC: 900

(Inst Memb) $482   MICROFORM: UMI
(Indv Memb) $361   LANGUAGE: English
INDEXED/ABSTRACTED: Math R, Sci Abstr
TARGET READER: Mathematicians interested in pure mathematics

Covers current research in all fields of pure mathematics.

STATISTICAL SCIENCE
ISSN: 0883-4237

EDITOR: Carl N. Morris
                                 Department of Statistics
                                 Science Center
                                 Harvard University
                                 One Oxford Street
                                 Cambridge, MA 02138
MS REQUIREMENT: Four copies, dbl spaced, abstracts up to
200 words, 5 or 6 key words
PUBLISHER: Institute of Mathematical Statistics
DATE FOUNDED: 1986   FREQUENCY: Quarterly
PRICE: $60   CIRC: 3500
MICROFORM: UMI   LANGUAGE: English
INDEXED/ABSTRACTED: Qual Control Appl Stat, Stat Theory
Meth Abstr.
TARGET READER: Statisticians, Probabilists, Students and
Researchers of statistics and probability

Publishes discussions of methodological and theoretical topics of current
interest and importance, surveys of substantive research areas with
promising statistical applications, comprehensive book reviews,
discussions of classic articles from the statistical literature, and
interviews with distinguished statisticians and probabilists. Presents the
full range of contemporary statistical thought at a technical level
assessible to the broad community.

STATISTICS AND PROBABILITY LETTERS
ISSN: 0167-7152

EDITOR: Richard A. Johnson
                            Department of Statistics
                            1210 West Dayton Street
                            University of Wisconsin
                            Madison, WI 53706
MS REQUIREMENT: Triplicate, dbl spaced, abstract no more
than 40 words, key words and phrases, paper not to exceed six
printed pages
PUBLISHER: North-Holland
DATE FOUNDED: 1982   FREQUENCY: Monthly
PRICE: $421.34   LANGUAGE: English
INDEXED/ABSTRACTED: Cam Sci Abstr, Compumath, ISI Curr
Cont, J Cont Quant Meth, Math R, Zent Math
TARGET READER: Statisticians, Probabilists

An international letters journal covers all fields of statistics and
probability, and provides an outlet for rapid publication of short
communications in the field.

STOCHASTIC ANALYSIS AND APPLICATIONS
ISSN: 0736-2994

EDITOR: V. Lakshmikantham
                            Department of Applied Mathematics
                            Florida Institute of Technology
                            150 West University Blvd.
                            Melbourne, FL 32901-6988
MS REQUIREMENT: Original and two copies, 1½ line spaced,
100 word abstract
PUBLISHER: Marcel Dekker
DATE FOUNDED: 1983   FREQUENCY: Quarterly
PRICE: $310   LANGUAGE: English
MICROFORM: RPI
INDEXED/ABSTRACTED: Curr Ind Stat, ISI/Compumath, Math R,
Stat Theory Meth Abstr, Zent Math

TARGET READER: Statisticians, Probabilists

Provides for the rapid publication of papers covering any aspect of the development and application of stochastic analysis techniques in all areas of scientific endeavor.

STOCHASTIC PROCESSES AND THEIR APPLICATIONS
ISSN: 0304-4149

EDITOR:	P. Jagers
	School of Mathematical and
	Computing Sciences
	Chalmers University of Technology
	and Gothenburg University,
	S-41296
	Göteborg, SWEDEN

MS REQUIREMENT: Original and two copies, dbl spaced, abstract not to exceed 150 words, list of key words
PUBLISHER: North-Holland
DATE FOUNDED: 1973   FREQUENCY: 6/yr.
PRICE: $517.41MICROFORM: RPI, UMI
LANGUAGE: English, French
INDEXED/ABSTRACTED: Curr Ind Stat, ISI Curr Cont, Math R, Zent Math
TARGET READER: Mathematicians, Statisticians, Probabilists, Operations Researchers

An official journal of the Bernoulli Society for Mathematical Statistics and Probability. Publishes papers on the theory and application of stochastic processes. It is interested in papers dealing with characterization, structural properties, inference and control of stochastic processes.

STOCHASTICS AND STOCHASTICS REPORTS
ISSN: 1045-1129

HISTORY: *Stochastics*
                [ISSN: 0090-9491] V1-V25 (1973-1988)

EDITOR: Mark H. A. Davis
                        Department of Electrical Engineering
                        Imperial College
                        London SW7 2BT, ENGLAND
PUBLISHER: Gordon and Breach
DATE FOUNDED: 1989   FREQUENCY: Monthly
PRICE: Varies-Send   LANGUAGE: English
        Request to Publisher   (mainly), French, German
MICROFORM: MIM
INDEXED/ABSTRACTED: Math R, Sci Abstr
TARGET READER: Applied Mathematicians, Engineers, Life
Scientists, Economists

Publishes articles dealing with all aspects of stochastic systems, with
related questions in the theory of stochastic processes and with
significant applications of stochastic process theory to problems in
engineering systems, the physical and life sciences, economics, and
other areas.

STUDIA LOGICA
ISSN: 0039-3215

EDITOR: Ryszard Wojcicki
                        90-954 Lodz 4
                        P. O. Box 79
                        POLAND
MS REQUIREMENT: Triplicate, brief abstract
PUBLISHER: Polish Academy of Sciences
DIST BY: Kluwer Academic Publishers
DATE FOUNDED: 1953 FREQUENCY: Quarterly
PRICE: (Inst) $160.50LANGUAGE: English
        (Indv) $65.50
INDEXED/ABSTRACTED: Linguist Lang Behav Abstr, Math R,
Phil Ind, Ref Zh, Zent Math
TARGET READER: Philosphers, Logicians, Linguists

Publishes papers on all technical issues of contemporary logics as well
as on philosophical issues, especially in linguisitics and other sciences
which depend on logic concepts, methods, and results. Interested in
papers on logical systems, their semantics, methodology, and
application.

STUDIA MATHEMATICA
ISSN: 0039-3223

SEND MS TO: Studia Mathematica
                       Sniadeckich 8
                       P. O. Box 137
                       00-950 Warszawa,
                       POLAND
                       TELEX 816112 PANIM PL
MS REQUIREMENT: Original and one copy, dbl spaced,
abstract not to exceed 200 words, AMS Subject Classification
Numbers
PUBLISHER: Institute of Mathematics
                       Polish Academy of Science
SEND REQUEST TO: Publisher
DATE FOUNDED: 1929   FREQUENCY: Irr
PRICE: Varies LANGUAGE: English, French,
                                       German, Russian
INDEXED/ABSTRACTED: Math R
TARGET READER: Analysts, Probabilists

Publishes original papers in functional analysis, abstract methods of
mathematical analysis and probability theory.

STUDIES IN APPLIED MATHEMATICS
ISSN: 0022-2526

HISTORY: *Journal of Mathematics and Physics*
                [ISSN: 0097-1421] V1-V47 (1921-1968)

EDITOR: David J. Benney

Department of Math,
MIT, Rm 2-341
Cambridge, MA 02139
PUBLISHER: Elsevier
DATE FOUNDED: 1969 FREQUENCY: Monthly
PRICE: $248 MICROFORM: UMI
LANGUAGE: English
INDEXED/ABSTRACTED: Appl Mech Rev, Curr Cont, Eng Ind,
Math R, Sci Cit Ind
TARGET READER: Applied Mathematicians, Engineers,
Geophysicists, Physicists, Meteorologists.

Publishes research papers involving the core concepts of applied
mathematics research, the major fields covered are computer science,
mechanics, astrophysics, geophysics, statistics, and probability.

SUGAKU EXPOSITIONS
ISSN: 0898-9583

English translation of: *Sugaku*

PUBLISHER: American Mathematical Society
DATE FOUNDED: 1988 FREQUENCY: Semi-Annually
PRICE: (Inst) $81   CIRC: 950
       (Inst Memb) $65   LANGUAGE: English
       (Indv Memb) $49
INDEXED/ABSTRACTED: Math R
TARGET READER: Mathematicians in all fields

Consists of translations of the expository articles in Sugaku and is
published by arrangement with Iwanami Shoten Publishers, Tokyo, and
the Mathematical Society of Japan.   Provides highly informative
accounts of a variety of current areas of research.

SURVEYS ON MATHEMATICS FOR INDUSTRY*
ISSN: 0938-1953

MANAGING
EDITOR: H. Engl
PUBLISHER: Springer-Verlag
DATE FOUNDED: 1991   FREQUENCY: Quarterly
PRICE: $169 LANGUAGE: English
TARGET READER: Applied Mathematicians, Industry
Engineers

Publishes original articles on the latest mathematical techniques relevant
for industry as well as exposing industrial problems of interest to
mathematicians.

*SYMPOSIA MATHEMATICA**
ISSN: 0082-0725

PUBLISHER: Academic Press
DATE FOUNDED: 1969   FREQUENCY: Irr
PRICE: $95   REPRINT: ISI
LANGUAGE: English, French, German, Italian
INDEXED/ABSTRACTED: Math R, Zent Math
TARGET READER: Mathematicians in all fields

Publishes original mathematical papers in pure and applied fields.

THEORETICAL AND MATHEMATICAL PHYSICS
ISSN: 0040-5779

Translation of *Theoreticheskaya i Matematicheskaya Fizika*
        [ISSN: 0564-6162]

EDITOR: A. A. Logunov
PUBLISHER: Consultant Bureau
DATE FOUNDED: 1969   FREQUENCY: 4/yr.
PRICE: $945 LANGUAGE: English
INDEXED/ABSTRACTED: Appl Mech Rev, Curr Cont, Math R,
Zent Math

TARGET READER: Mathematicians, Theoretical Physicists

Reports on current developments in theoretical physics as well as mathematical problems related to theoretical physics.

THEORY OF PROBABILITY AND ITS APPLICATIONS
ISSN: 0040-585X

Translation of the Russian journal
    Teoriya Veroyathnosteiiee Primeneniya

EDITOR:	N. Brunswick, Retired Mathematician
	B. Seckler
	C. W. Post Center
	Long Island University
	Greenvale, NY 11548
PUBLISHER:	SIAM

DATE FOUNDED:	1956	FREQUENCY:	Quarterly
PRICE: (Inst)	$300	CIRC:	1200
(Indv)	$99	LANGUAGE:	English

Back Issues Available
INDEXED/ABSTRACTED: Appl Mech Rev, Compumath, Comput Rev, Curr Cont, Math R, Sci Abstr
TARGET READER: Applied Mathematicians, Probabilists, Statisticians

Contains papers on the theory and application of probability, statistics, and stochastic processes.

THEORY OF PROBABILITY AND MATHEMATICAL STATISTICS
ISSN: 0094-9000

TRANSLATION
EDITOR:	Simeon Ivanov
PUBLISHER:	American Mathematical Society
DATE FOUNDED:	1974 FREQUENCY: Semi-Annually

PRICE: (Inst) $275   CIRC: 2500
                  (Inst Memb) $220   LANGUAGE: English
INDEXED/ABSTRACTED: J Cont Quant Meth, Math R
TARGET READER: Probabilists, Statisticians

Cover to cover translation of Teoriya Veroyatnoslei i Matematicheskaya
Statistika. Originally published in Russian by Kiev University.

TOHOKU MATHEMATICAL JOURNAL
ISSN: 0040-8735

SEND MS TO: Editor, Tohoku Mathematical Journal
                         Mathematical Institute
                         Tohoku University
                         Sendai, 980, JAPAN
MS REQUIREMENT: Typewritten, dbl spaced
PUBLISHER: Mathematical Institute of Tohoku
                         University
                         Maruzen Co., Ltd.
DATE FOUNDED: 1911   FREQUENCY: Quarterly
        (Series 2)   1949   PRICE: $187
LANGUAGE: English, French, German
LIBRARY EXCHANGE: Exchanges should be sent to the
Librarian of the Tohoku Mathematical Journal at the Mathematical
Institute, Tohoku University, Sendai, 980, Japan.
INDEXED/ABSTRACTED: Math R
TARGET READER: Mathematicians in all fields

Publishes high academic level research papers in both pure and applied
mathematics.

TOPOLOGY
ISSN: 0040-9383

EDITOR: Ralph Cohen
                         Stanford University
                         Stanford, CA 94305

MS REQUIREMENT: Duplicate
PUBLISHER: Pergamon Press
DATE FOUNDED: 1962 FREQUENCY: 6/yr.
PRICE: $380 MICROFORM: UMI, Pergamon
LANGUAGE: English, French, German, Italian
INDEXED/ABSTRACTED: Compumath, Curr Cont, Math R
TARGET READER: Mathematicians, Topologists

Publishes papers in all areas of mathematics, but with special emphasis on subjects which are related to topology or geometry.

TOPOLOGY AND ITS APPLICATIONS
ISSN: 0166-8641

HISTORY: *General Topology and its Application*
   [ISSN: 0016-660X] V1-V10 (1971-1979)

EDITOR: Richard B. Sher
   Department of Math
   University of No. Carolina
   Greensboro, NC 27412
PUBLISHER: North-Holland
DATE FOUNDED: 1980 FREQUENCY: 15/yr.
PRICE: $814.60 CIRC: 700
MICROFORM: RPI LANGUAGE: English
INDEXED/ABSTRACTED: ISI Curr Cont, Math R, Zent Math
TARGET READER: Topologists (Set Theoretic as well as Algebraic)

Devoted to broad coverage of the various areas of topological research. The subjects include the algebraic, general, geometric, and set-theoretic facets of topology as well as areas of interactions between topology and other mathematical disciplines.

TOPOLOGY PROCEEDINGS
ISSN: 0146-4124

SEND MS TO:	Donna Bennett, Technical Editor
	Topology Proceedings
	Mathematics F.A.T. Department
	Auburn University
	Auburn, AL 36849
MS REQUIREMENT:	Dbl spaced, E-Mail manuscripts written in
TEX are encouraged	
PUBLISHER:	Auburn Univeristy,
	Math F.A.T. Department
DATE FOUNDED:	1976 FREQUENCY: Semi-Annually
PRICE:	$55 CIRC: 300
LANGUAGE:	English PUBL CHRG: $25/Page
	Paid by Institution

INDEXED/ABSTRACTED: Math R
TARGET READER: Topologists

Publishes the proceedings of the Annual Topology Conference. Each volume consists of three distinct parts: Section I contains refereed research and survey articles, Section II contains research announcements, and Section III is devoted to unsolved problems in topology.

TRANSACTIONS OF THE AMERICAN MATHEMATICAL SOCIETY
ISSN: 0002-9947

MANAGING	David J. Saltman,
EDITOR:	Department of Math
	University of Texas/Austin
	Austin, TX 78713
MS REQUIREMENT:	Duplicate, dbl spaced, accept papers of
electronic form	
PUBLISHER:	American Mathematical Society
DATE FOUNDED:	1900 FREQUENCY: Monthly
PRICE: (Inst)	$780 MICROFORM: UMI

(Inst Memb) $624   CIRC: 1700
LANGUAGE: English
INDEXED/ABSTRACTED: Compumath, Curr Cont, Eng Ind,
Math R
TARGET READER: Mathematicians in all fields.

Devoted entirely to research in pure and applied mathematics, and, in general, includes longer papers than those in the Proceedings. Papers should be at least 16 typed pages in length.

TRANSACTIONS OF THE MOSCOW MATHEMATICAL SOCIETY
ISSN: 0077-1554

English Edition of: *Moskovskogo Matematscheskogo Obshchestva.*
     Trudy

PUBLISHER: American Mathematical Society
DATE FOUNDED: 1978   FREQUENCY: Annually
PRICE: (Inst) $193   CIRC: 500
         (Inst Memb) $154   LANGUAGE: English
INDEXED/ABSTRACTED:Math R
TARGET READER: Mathematicians in the field of pure mathematics.

Journal is the translation of *Trudy Moskovskogo Matematicheskogo Obshchestva* and contains the results of original research in pure mathematics.

UKRAINIAN MATHEMATICAL JOURNAL
ISSN: 0041-5995

English Translation of: *Ukrainskii Matematicheskii Zhurnal*
     [ISSN: 0041-6053]

PUBLISHER: Consultant Bureau
DATE FOUNDED: 1967   FREQUENCY: Bi-Monthly

PRICE: $995   LANGUAGE: English
INDEXED/ABSTRACTED: Comput Inf Syst Abstr J, Math R,
Zent Math
TARGET READER: Mathematicians in all fields

Publishes articles in various areas of pure and applied mathematics.

UMAP JOURNAL
(Undergraduate Mathematics Applications Project) Journal
ISSN: 0197-3622

EDITOR: Paul J. Campbell
                          Beloit College
                          700 College Street
                          Beloit, WI 53511
PUBLISHER: Consortium for Mathematics
                          and its Applications
DATE FOUNDED: 1980   FREQUENCY: Quarterly
PRICE: (Inst) $148   CIRC: 2000
               (Indv) $51   LANGUAGE: English
               (Lbry) $98
INDEXED/ABSTRACTED: Educ Ind
TARGET READER: Mathematics Educators, Undergraduate
Math Students

Publishes papers with a wide variety of professional applications of the
mathematical sciences and provides a forum for the discussion of new
directions on mathematical education.

USSR COMPUTATIONAL MATHEMATICS AND MATHEMATICAL PHYSICS
ISSN: 0041-5553

English translation of: *Zhurnal Vychislitel'noi Matematiki i
Matematicheskoi Fiziki*

SCIENTIFIC   R. C. Glass

	City University	
	Northampton Square	
	London, EC1V 0HB, ENGLAND	
PUBLISHER	Pergamon Press	
DATE FOUNDED:	1962 FREQUENCY:	Bi-Monthly
PRICE:	$1295 CIRC:	1000
LANGUAGE:	English MICROFILM:	Pergamon, MIM, UMI

INDEXED/ABSTRACTED: Cam Sci Abstr, Compumath, Curr Cont, Math R, Sci Search
TARGEST READER: Scientists, Mathematicians, Physicists

Publishes papers on mathematical problems arising in all fields of science, e.g., physics, mechanics, computer theory, as well as purely mathematical nature.

1991-1992 NEW MATHEMATICAL JOURNAL TITLES

TABLE OF ABSTRACTS AND INDEXES

Linguist Abstr	Linguistics Abstracts
Linguist Lang Behav Abstr	Linguistics and Language Behavior Abstracts
Manage Cont	Management Contents
Math R	Mathematical Reviews
Oper Res Manag Sci	Operations Research Management Science
Phil Ind	Philosopher's Index
Phys Abstr	Physics Abstracts
Pollut Abstr Ind	Pollution Abstracts with Indexes
Popul Ind	Population Index
Psychol Abstr	Psychological Abstracts
Qual Control Appl Stat	Quality Control and Applied Statistics
Ref Zh	Referativnyi Zhurnal
Ref Zh Biol	Referativnyi Zhurnal. Biologiia
Ref Zh Fizika	Referativnyi Zhurnal. Fizika
Risk Abstr	Risk Abstracts
SCI	Social Sciences Index
Sci Abstr	Science Abstracts
Sci Cit Ind	Science Citation Index
Soc Sci Ind	Social Sciences Index
Socio Abstr	Sociological Abstracts
SSCI	Social Science Citation Index
Stat Ref Ind	Statistical Reference Indexes
Stat Theory Meth Abstr	Statistical Theory and Methods Abstracts
Zent Math	Zentralblatt fur Mathematik und ihre Grenzgebiete

DIRECTORY OF PUBLISHERS AND DISTRIBUTORS

Directory DOES NOT list complete addresses of all publishers and distributors. The following addresses are for ordering the journals compiled in this book only.

Ablex Publishing Corporation
355 Chestnut Street
Norwood, NJ 07648
Tel: (201) 767-8450

Academic Press, Inc.
1 East First Street
Duluth, MN 55802

or

Journal Promotion Department
1250 Sixth Avenue
San Diego, CA 92101
Tel: (619) 699-6742

or

24-28 Oval Road
London NW1 7DX, ENGLAND
Tel: (071) 267-4466

or

Foots Cray
Sidcup, Kent, DA14 5HP, ENGLAND
Tel: (01) 300-3322

Akademiai Kiado
Publishing House of the Hungarian
Academy of Sciences
Orders may be placed with:
Kultura Foreign Trading Co.
H-1389, Budapest
P. O. Box 149
Budapest, HUNGARY

Akademie Verlag
GmbH, Leipzipen Str. 3-4
Postfach 1233
0-1086 Berlin, GERMANY
Tel: 22366350
TELEX: 114420

Allen Press
      1041 New Hampshire Street
      P. O. Box 1897
      Lawrence, KS 66044-8897
      Tel: (913) 843-1221
      FAX: (913) 843-1274
Allerton Press, Inc.
      150 Fifth Avenue
      New York, NY 10011
      Tel: (212) 924-3950
Almqvist & Wiksell International
      P. O. Box 638
      S-101 28 Stockholm, SWEDEN
American Institute of Physics
      500 Sunnyside Blvd.
      Woodbury, NY 11797-2999
American Mathematical Society
      Box 1571, Annex Station
      Providence, RI 02901-9930
      Tel: (401) 455-4000
      Toll Free: 1-800-321-4267
      (charge orders only)
American Sciences Press, Inc.
      20 Cross Road
      Syracuse, NY 13224-2144
American Statistical Association
      1429 Duke Street
      Alexandria, VA 22314
Amministrazione Degli Annali
      Classe de Scienze
      Scuola Normale Superiore di Pisa
      Piazza dei Cavalieri
      7-56100, Pisa, ITALY
      Telex: (590) 548-SNSPI
      Telefax: 563-513
Applied Probability Trust
      Dept. of Probability and Statistics
      The University
      Sheffield, S3 7RH, ENGLAND

Association for Symbolic Logic
Journal Division
University of Illinois
Department of Mathematics
1409 West Green Street
Urbana, IL 61801
Auburn University
Mathematics F.A.T. Department
Auburn, AL 36849
Auckland Mathematical Association, Inc.
P. O. Box 26-226
Auckland 3, NEW ZEALAND
Australian Mathematical Society
Department of Math
University of Queensland
Queensland, 4072, AUSTRALIA
or
School of Mathematical and Physical Sciences
Murdoch University
Perth, WA 6150
Australian Statistical Publishing Association, Inc.
G. P. O. Box 573
Canberra
A.C.T. 2601, AUSTRALIA
B. G. Teubner
Industriestr. 15
Postfach 801069
D-7000 Stuttgart 80, GERMANY
Basil Blackwell, Ltd.
3 Cambridge Center
Cambridge, MA 02142
or
108 Cowley Road
Oxford, OX4 1JF, ENGLAND
Baywood Publishing Co., Inc.
26 Austin Avenue
P. O. Box 337
Amityville, NY 11701
Tel: (516) 691-1270

Bibliographisches Institut & F.A.
    Brockhaus A. G.
    Postfach 10 03 11
    D-6800
    Mannheim 1, GERMANY
Biometric Society
    1429 Duke St., Suite 401
    Alexandria, VA 22314
Birkhaüser-Verlag
    P. O. Box 133, CH-4010
    Basel, SWITZERLAND
    Tel: (061) 737740
British Psychological Society
    St. Andrews House
    48 Princess Rd., East
    Leicester, LE1 7DR, ENGLAND
Brown University
    Providence, RI 02912
Butterworths
    Journal Promotion Department
    80 Montvale Avenue
    Stoneham, MA 02180
    Tel: (617) 438-8464
    Toll Free: 1-800-366-2665
or
Butterworth Scientific Ltd.
    P. O. Box 63
    Westbury House
    Bury Street
    Guildford Surrey, GU2 5BH, ENGLAND
Calcutta Mathematical Society
    92 Acharya Prafulla Chandra Road
    Calcutta, INDIA 700009
Calcutta Statistical Association
    Cambridge University Press
    40 West 20th Street
    New York, NY 10011-4211
    Tel: (212) 924-3900
    Toll Free: 1-800-221-4512

or

    The Edinburgh Building
    Shaftesbury Road
    Cambridge, CB2 2RU, ENGLAND

Canadian Mathematical Society
    577 King Edward
    Ottawa, Ontario, CANADA K1N 6N5

Carfax Publishing Company
    85 Ash Street
    Hopkinton, MA 01748

or

    P. O. Box 25
    Abingdon, Oxfordshire
    OX 14 3UE, ENGLAND

Charles Babbage Research Center
    P. O. Box 512
    Pembina, ND 58271-0512

or

    P. O. Box 272
    St. Norbert Postal Station
    Winnipeg, Manitoba, CANADA R3V 1L6

CHZ Ars Polona
    P. O. Box 1001
    00-950 Warsaw, POLAND

**Consortium for Mathematics and
Its Applications**
    60 Lowell Street
    Arlington, MA 02174

Descartes Press Company
    2-13-10 Kaguike
    Koriyama
    Fukushima-ken, JAPAN 963
    Tel#:   0249-22-7596

Duke University Press
    6697 College Station
    Durham, NC 27708
    Tel: (919) 684-2173

Elsevier Science Publishers
>655 Avenue of the Americas
>New York, NY 10010
>Tel: (212) 633-3950
>FAX: (212) 633-3990

or

>Journal Division
>P. O. Box 211, 1000 AE
>Amsterdam, THE NETHERLANDS

L'Enseignement Mathématique
>Case Postale 240
>1211 Genève 24, SUISSE

Fibonacci Association
>Santa Clara University
>Santa Clara, CA 95053

Forum for Interdisciplinary Mathematics
>F-9-12 Model Town
>Delhi, 110009, INDIA

Gauthier-Villars, North American, Inc.
>American Promotion Office
>875-81 Massachusetts Avenue
>Cambridge, MA 02139
>Tel: (617) 354-7875
>FAX: (617) 354-6875

or

Gauthier-Villars
>Journal's Department
>15, Rue Gossin
>92543 Montrouge Cedex, FRANCE
>Tel: (33-1) 40-92-65-00
>Telefax: (33-1) 40.92.65.97

Gordon and Breach Science Publisher
>P. O. Box 786 Cooper Station
>New York, NY 10276
>Tel: (212) 206-8795
>FAX: (212) 645-2459
>TELEX: 236735 GOPUB UR

or

>P. O. Box 197
>London, WC2E 9PX, ENGLAND

Guilford College
> Department of Math
> Greensboro, NC 27410
> Tel: (919) 292-5511

Hardronic Press, Inc.
> 35246 US 19, No., #131
> Palm Harbor, FL 34683

Harwood Academic Publisher
> 270 Eighth Avenue
> New York, NY 10011
> Tel: (212) 206-8900

Houston Journal of Mathematics
> University of Houston
> Houston, TX 77204-5883

Indian Academy of Sciences
> CV Raman Avenue
> P. O. Box 8005
> Bangalore, 560 080, INDIA

Indian Institute of Technology
> Madras
> Humanities and Sciences Building
> Room HsB 249
> First Floor
> Madras, 600 036, INDIA

Indian National Science Academy
> Bahadur Shah Zafar Marg
> New Delhi, 110002, INDIA

Indiana University Mathematics Journal
> Swain Hall East 222
> Bloomington, IN 47405

Industrial Mathematics Society
> P. O. Box 159
> Roseville, MI 48066

Institut Mittag-Leffler
> Auravägen 17, S-18262
> Djursholm, SWEDEN

Institute of Management Sciences
> 290 Westminster Street
> Providence, RI 02903

Institute of Mathematical Statistics
        3401 Investment Blvd., #7
        Hayward, CA 94545
        Tel: (415) 783-8141
International Journal of Mathematics
and Mathematical Sciences
        University of Central Florida
        Orlando, FL 32816
International Statistical Institute
        Subscription Department
        428 Prinses Beatrixlaan
        P. O. Box 950
        2270 AZ Voorburg, NETHERLANDS
IOP Publishing, Ltd.
        Techno House, Redcliffe Way
        Bristol BS1 6NX, ENGLAND
        Tel: 0272-297481
        FAX: 0272-294318
        TELEX: 449149

ISI
        Institute for Scientific Information
        University City Science Center
        3501 Market Street
        Philadelphia, PA 19104
        FAX: (215) 386-6362
        TELEX: 845305
J. C. Baltzer AG
Scientific Publishing Co.
        Wettsteinplatz 10
        CH-4058, Basel, SWITZERLAND
J. Fang
        P. O. Box 206
        Woods Cross Roads, VA 23190
J. W. Arrowsmith, Ltd.
        71 Winterstoke Road
        Bristol, BS3 2NT, ENGLAND
Japanese Association of Mathematical Sciences
        Shin Sakaihigashi Bldg.
        2-1-18 Minami Hanadaguchi
        Sakai, Osaka, 590, JAPAN

John Wiley & Sons, Ltd.
 Subscription Department
 605 Third Avenue
 New York, NY 10158

or

 Baffins Lane, Chichester
 Sussex, PO19 1UD, ENGLAND

Johns Hopkins University Press
 Journals Publishing Division
 701 West 40th St., Suite 275
 Baltimore, MD 21211

Johnson Reprint Corporation
 111 Fifth Avenue
 New York, NY 10003

Karl Weierstrass-Institut für Mathematik der AdW
 Mohrenstrasse 39
 DDR-1086, Berlin, GERMANY
 Tel: 20 37 73 03

Kinokuniya Co., Ltd.
 17-7 Shinjuku 3 chome, Shinjuku-ku
 Tokyo, 106, JAPAN

Kluwer Academic Publishers
 101 Philip Drive
 Assinippi Park
 Norwell, MA 02061

or

 P. O. Box 358
 Accord Station
 Hingham, MA 02018-0358

or

 P. O. Box 322
 3300 AA Dordrecht
 THE NETHERLANDS

Kobe University
 College of Liberal Arts
 Department of Mathematics
 2-1, Tsurukabuto 1-chome
 Nada-ku, Kobe-shi
 Hyogo-ken 657, JAPAN

Korean Mathematical Society
538 Dowha Dong
Mapo Ku
Sung Ji Building, Room 706
Seoul 121-743, KOREA

KTO
Kraus Microform
Route 100
Millwood, NY   10546
Tel: (914) 762-2200
FAX: (914) 762-1195
TELEX: 6818112

Kultura Foreign Trade Enterprise
Periodicals Export Department
Fo Utca 32, P. O. Box 149
1389 Budapest, HUNGARY

Kumamoto University
Kumamoto 860, JAPAN

Lehigh University
Christmas-Saucon Hall 14
Bethlehem, PA   18015-3125

London Mathematical Society
Distribution Center
Blackhorse Road
Letchworth, Herts SG6 1HN, ENGLAND

or

Burlington House
Piccadilly, London
W1V 0NL, ENGLAND

Marcel Dekker, Inc.
270 Madison Avenue
New York, NY   10016
Tel: (212) 796-9000
FAX: (914) 976-1772

Maruzen Co., Ltd.
P. O. Box 5050
Tokyo International
100-31, JAPAN
Tel: 031278-9224
FAX: 81-3-274-2270

TELEX: 026517 Maruzen
Matematisk Institut
    NY Munkegade
    8000 Arhus C, DANMARK
Mathematical Association
    259 London Road
    Leicester, LE2 3BE, ENGLAND
Mathematical Association of America
    1529 18th St., NW
    Washington, D.C. 20036
    Tel: (202) 387-5200
Mathematical Society of Japan
    25-9-203, Hongo 4-chome,
    Bunkyo-ku
    Tokyo 113, JAPAN
MATYC Journal, Inc.
    P. O. Box 158
    Old Beth Page, NY 11804
Michigan State University
    Department of Math
    East Lansing, MI 48824-1027
MIM
    Microforms International
    Maxwell House
    Fairview Park
    Elmsford, NY 10523
    Tel: (914) 592-7700
    FAX: (914) 592-3625
    TELEX: 13-7328
Minnesota State University
    Department of Math
    St. Olaf College
    Northfield, MN 55057
Myron E. Sharpe, Inc.
    80 Business Park Drive
    Armonk, NY 10504
National Academy of Sciences
    2102 Constitution Avenue
    Washington, D.C. 20418
    Tel: (202) 334-2525

Toll Free: 1-800-624-6242
National Council of Teachers of Mathematics
        1906 Association Drive
        Reston, VA 22091
        Tel: (703) 620-9840
Nihon Tokei Gakkai
        Japan Statistical Society
        c/o Institute of Statistical Mathematics
        4-6-7, Minami-Azabu
        Minato-Ku, Tokyo 106, JAPAN
Notre Dame Journal of Formal Logic
        Box 5
        Notre Dame, IN 46556
        Tel: (219) 239-6157
Nova Science Publishers, Inc.
        283 Commack Rd., Suite 300
        Commack, NY 11725-3401
        Tel: (516) 499-3103
Ohio University Press
        Scott Onadrangle
        Athens, OH 45701
        Tel: (614) 593-1155
        Toll Free: 1-800-666-2211
Okayama University
        Faculty of Science
        Department of Mathematics
        3-1-1 Tsushima-Naka
        Okayama-shi, Okayama-ken 700
        JAPAN
Operations Research Society of America
        Mount Royal and Guilford Avenues
        Baltimore, MD 21202
        Tel: (301) 528-4146
Oxford University Press
        200 Madison Avenue
        New York, NY 10016
        Tel: (212) 679-7300

or

Journal Subscription Department
Pinkhill House, Southfield Road
Eynsham, Oxford, OX8 1JJ, ENGLAND

Pacific Journal of Mathematics
P. O. Box 969
Carmel Valley, CA 93924

Panamerican Mathematical Journal
University of Central Florida
Department of Math
Orlando, FL 32816

Pergamon Press, Inc.
Journal Division
Maxwell House, Fairview Park
Elmsford, NY 10523
Tel: (914) 592-7700
Toll Free: 1-800-257-5755

Pi Mu Epsilon
Macalester College
St. Paul, MN 55105
Tel: (612) 696-6057

Pitagora Editrice
c/o Paola Liberti
Instituto di Matematica Applicata
"G. Sansone"
Facolta di Ingegneria
Universita di Firenze
Via S. Marta 3
50139 Firenze, ITALY

Plenum Publishing Corporation
233 Spring Street
New York, NY 11003
Tel: (212) 620-8000
Toll Free: 1-800-221-9369
FAX: (212) 463-0742
TELEX: 23/421139

Polish Academy of Sciences
      Institute of Mathematics
      P. O. Box 137
      00-950 Warszawa, POLAND
      TELEX: 816112 PANIM PL
Princeton University Press
      41 William Street
      Princeton, NJ 08540
      Tel: (609) 258-4900
      Toll Free: 1-800-777-4726
R. Oldenbourg Verlag GmbH
      Rosenheimer Str., 145
      Postfach 80, 1360
      8000 Munich 80, GERMANY
Research Association of Statistical Sciences
      Kyushu University 33
      Fukuoka 812, JAPAN
Rocky Mountain Mathematics Consortium
      Department of Math
      Arizona State University
      Tempe, AZ 85287
Rose-Hulman Institute of Technology
      Terre Haute, IN 47803
Royal Society of Canada
      Department of Pure Math
      University of Waterloo
      Waterloo, Ontario, N2L 3G1, CANADA
Royal Society of Edinburgh
      22 George Street
      Edinburgh, EH2 2PQ, SCOTLAND
RPI
      Research Publications, Inc.
      12 Lunar Drive, Drawer AB
      Woodbridge, CT 09525
      Tel: (203) 397-2600
      Toll Free: 1-800-732-2477
      FAX: (203) 397-3893
      TELEX: 710-4656345

Science Press
Beijing, CHINA
Scottish Academic Press, Ltd.
139 Leith Walk
Edinburgh, EH6 8NS, SCOTLAND
Tel: (031) 556-2796
SIAM
Society for Industrial and Applied Mathematics
University Science Center
Philadelphia, PA 19104-2688
Slovenska Akademia Vied
Publishing House of the Slovak
Academy of Sciences
Klemensova 19, 814 30 Bratislava
CZECHOSLOVAKIA
Sociedade Brasileira de Matemática
Aplicada E Computacional
Rua Lauro Müller, 455
22290-Rio de Janeiro, BRASIL
Tel: 541-2132
Springer-Verlag
Journal Fulfillment Services
P. O. Box 2485
Secaucus, NJ 07096-2491
Tel: (201) 348-4033
or
Springer-Verlag New York, Inc.
175 Fifth Avenue·
New York, NY 10010
Tel: (212) 460-1500
TELEX: 23 22 35
Stichting Mathematisch Centrum
Kruislaan 413
P. O. Box 4079
1009 AB Amsterdam, THE NETHERLANDS
Swets & Zeitlinger B.V.
Heereweg 347B
2160 Lisse, THE NETHERLANDS

Taylor & Francis, Ltd.
4 John Street
London, WC1N 2ET, ENGLAND
Tokushima University
1-1, Minami-Josanjima-cho
Tokushima-shi, Toshima-ken 700
JAPAN
Tokyo International
Import and Export
P. O. Box 5050
Tokyo 100-31, JAPAN
Toyama University
Department of Mathematics
Faculty of Science
Toyama, JAPAN
UMI
University Microfilm International
300 No. Zeeb Road
Ann Arbor, MI 48106
Tel: (313) 761-4700
Toll Free: 1-800-521-0600
                   1-800-521-3042
FAX: (313) 761-1203
TELEX: 211607
University of Auckland
Mathematical Chronicle Committee
Department of Mathematics and Statistics
Private Bag
Auckland, NEW ZEALAND
University of Houston
Houston, TX 77204-5883
University of Illinois Press
54 East Gregory Drive
Champaign, IL 61820
Tel: (217) 333-0950
Toll Free: 1-800-545-4703
University of Michigan
Department of Math
Ann Arbor, MI 48109-1003
(313) 764-0337

University of Notre Dame
P. O. Box L
Notre Dame, IN 46556
Tel: (219) 239-6346
Toll Free: 1-800-677-3232
University of Pittsburgh
Campus Drive
Bradford, PA 16701
University of the West Indies
Cave Hill, BARBADOS
University of Tokyo
Department of Mathematics
Faculty of Science
Bunkyo-ku
Tokyo 113, JAPAN
University of Toronto
Department of Math
Toronto, Ontario
M5S 1A1, CANADA

or

Journal Department Press
5201 Dufferin Street
Downsview, Ontario, CANADA M3H 5T8
VSP
P. O. Box 346
3700 AH Zeist
THE NETHERLANDS
Tel: 03404-51902
Walter de Gruyter & Co.
200 Saw Mill River Road
Hawthorne, NY 10532
Tel: (914) 747-0110
TELEX: 646677
FAX: (914) 747-1326

or

P. O. Box 11 02 40
D-1000, Berlin, GERMANY
Tel: (030) 26005-210
TELEX: 184027
FAX: (030) 26005

Weizmann Science Press of Israel
     P. O. Box 801
     Jerusalem, 91007, ISRAEL
World Scientific Publishing, Co., PTe Ltd.
     687 Hatrwell Street
     Teaneck, NJ 07666

or

     P.O. Box 128
     Farrer Road
     Singapore 9128, SINGAPORE
     Tel: 278-6188

INDEX OF INTERNATIONAL
STANDARD SERIAL NUMBERS

The numbers listed below are for those journals mentioned in this book. Any number with an asterisk (*) indicates an out-of-publication journal or a journal that was merged with another to create the current publication and is only listed as a reference guide.

ABOUT THE AUTHOR

DIANA F. LIANG (MA, Graduate degree in Library Science, Vanderbilt University) is currently a University Librarian at the University of South Florida Library. She is a cataloguer as well as a Collection Development Librarian responsible for the College of Natural Sciences, which includes the Mathematics Department. Prior to her current academic post, Ms. Liang worked as a County Librarian in Lancaster, Ohio, and as a Special Librarian at the Chemical Abstract Services in Columbus, Ohio. She is a longtime ALA member and serves on various committees in the library and in the university community. She has received numerous Professional Development Leaves and was a recipient of the University President's Council Grant award.